T0257944

New Frontiers in Plant In Vitro Culture

New Frontiers in Plant In Vitro Culture

Edited by **Clive Koelling**

New York

Published by Callisto Reference,
106 Park Avenue, Suite 200,
New York, NY 10016, USA
www.callistoreference.com

New Frontiers in Plant In Vitro Culture
Edited by Clive Koelling

© 2015 Callisto Reference

International Standard Book Number: 978-1-63239-478-1 (Hardback)

This book contains information obtained from authentic and highly regarded sources. Copyright for all individual chapters remain with the respective authors as indicated. A wide variety of references are listed. Permission and sources are indicated; for detailed attributions, please refer to the permissions page. Reasonable efforts have been made to publish reliable data and information, but the authors, editors and publisher cannot assume any responsibility for the validity of all materials or the consequences of their use.

The publisher's policy is to use permanent paper from mills that operate a sustainable forestry policy. Furthermore, the publisher ensures that the text paper and cover boards used have met acceptable environmental accreditation standards.

Trademark Notice: Registered trademark of products or corporate names are used only for explanation and identification without intent to infringe.

Printed in the United States of America.

Contents

Preface

Plant in vitro culture forms an essential part in the field of botany. This book serves the objective of presenting recent developments in the field of plant in vitro culture in relation with medicinal plants and perennial fruit crops. It extensively covers fundamental principles and recent techniques. It includes contributions of eminent researchers and experts in this field from all over the world. This book will serve as a valuable reference for students, teachers, researchers in biotechnology and for individuals interested in commercial applications of plant in vitro culture.

This book unites the global concepts and researches in an organized manner for a comprehensive understanding of the subject. It is a ripe text for all researchers, students, scientists or anyone else who is interested in acquiring a better knowledge of this dynamic field.

I extend my sincere thanks to the contributors for such eloquent research chapters. Finally, I thank my family for being a source of support and help.

Editor

Plant Tissue Culture:
Current Status and Opportunities

Altaf Hussain, Iqbal Ahmed Qarshi, Hummera Nazir and Ikram Ullah

Additional information is available at the end of the chapter

1. Introduction

Tissue culture is the *in vitro* aseptic culture of cells, tissues, organs or whole plant under controlled nutritional and environmental conditions [1] often to produce the clones of plants. The resultant clones are true-to type of the selected genotype. The controlled conditions provide the culture an environment conducive for their growth and multiplication. These conditions include proper supply of nutrients, pH medium, adequate temperature and proper gaseous and liquid environment.

Plant tissue culture technology is being widely used for large scale plant multiplication. Apart from their use as a tool of research, plant tissue culture techniques have in recent years, become of major industrial importance in the area of plant propagation, disease elimination, plant improvement and production of secondary metabolites. Small pieces of tissue (named explants) can be used to produce hundreds and thousands of plants in a continuous process. A single explant can be multiplied into several thousand plants in relatively short time period and space under controlled conditions, irrespective of the season and weather on a year round basis [2]. Endangered, threatened and rare species have successfully been grown and conserved by micropropagation because of high coefficient of multiplication and small demands on number of initial plants and space.

In addition, plant tissue culture is considered to be the most efficient technology for crop improvement by the production of somaclonal and gametoclonal variants. The micropropagation technology has a vast potential to produce plants of superior quality, isolation of useful variants in well-adapted high yielding genotypes with better disease resistance and stress tolerance capacities [3]. Certain type of callus cultures give rise to clones that have inheritable characteristics different from those of parent plants due to the possibility of occurrence of somaclonal variability [4], which leads to the development of commercially important improved varieties. Commercial production of plants through micropropagation techniques has several advantages over the traditional methods of propagation through seed, cutting, grafting and air-layering etc. It is rapid propagation processes that can lead to the production of plants virus free [5]. *Coryodalis yanhusuo*, an important medicinal plant was propagated by somatic embryogenesis from tuber-derived callus to produce disease free tubers [6]. Meristem tip culture of banana plants devoid from banana bunchy top virus (BBTV) and brome mosaic virus (BMV) were produced [7]. Higher yields have been obtained by culturing pathogen free germplasm *in vitro*. Increase in yield up to 150% of virus-free potatoes was obtained in controlled conditions [8]. The main objective of writing this chapter is to describe the tissue culture techniques, various developments, present and future trends and its application in various fields.

2. History of plant tissue culture

The science of plant tissue culture takes its roots from the discovery of cell followed by propounding of cell theory. In 1838, Schleiden and Schwann proposed that cell is the basic structural unit of all living organisms. They visualized that cell is capable of autonomy and therefore it should be possible for each cell if given an environment to regenerate into whole plant. Based on this premise, in 1902, a German physiologist, Gottlieb Haberlandt for the first time attempted to culture isolated single palisade cells from leaves in knop's salt solution enriched with sucrose. The cells remained alive for up to one month, increased in size, accumulated starch but failed to divide. Though he was unsuccessful but laid down the foundation of tissue culture technology for which he is regarded as the father of plant tissue culture. After that some of the landmark discoveries took place in tissue culture which are summarized as under:

- 1902 - Haberlandt proposed concept of *in vitro* cell culture
- 1904 - Hannig cultured embryos from several cruciferous species
- 1922 - Kolte and Robbins successfully cultured root and stem tips respectively
- 1926 - Went discovered first plant growth hormone –Indole acetic acid
- 1934 - White introduced vitamin B as growth supplement in tissue culture media for tomato root tip
- 1939 - Gautheret, White and Nobecourt established endless proliferation of callus cultures
- 1941 - Overbeek was first to add coconut milk for cell division in *Datura*
- 1946 - Ball raised whole plants of *Lupinus* by shoot tip culture
- 1954 - Muir was first to break callus tissues into single cells
- 1955 - Skoog and Miller discovered kinetin as cell division hormone
- 1957 - Skoog and Miller gave concept of hormonal control (auxin: cytokinin) of organ formation
- 1959 - Reinert and Steward regenerated embryos from callus clumps and cell suspension of carrot (*Daucus carota*)
- 1960 - Cocking was first to isolate protoplast by enzymatic degradation of cell wall

- 1960 - Bergmann filtered cell suspension and isolated single cells by plating
- 1960 - Kanta and Maheshwari developed test tube fertilization technique
- 1962 - Murashige and Skoog developed MS medium with higher salt concentration
- 1964 - Guha and Maheshwari produced first haploid plants from pollen grains of *Datura* (Anther culture)
- 1966 - Steward demonstrated totipotency by regenerating carrot plants from single cells of tomato
- 1970 - Power *et al.* successfully achieved protoplast fusion
- 1971 - Takebe *et al.*regenerated first plants from protoplasts
- 1972 - Carlson produced first interspecific hybrid of *Nicotiana tabacum* by   protoplast fusion
- 1974 - Reinhardintroduced biotransformation in plant tissue cultures
- 1977 - Chilton *et al.* successfully integrated Ti plasmid DNA from *Agrobacterium tumefaciens* in plants
- 1978- Melchers *et al.* carried out somatic hybridization of tomato and potato resulting in pomato
- 1981- Larkin and Scowcroft introduced the term somaclonal variation
- 1983 - Pelletier *et al.*conducted intergeneric cytoplasmic hybridization in Radish and Grape
- 1984 - Horsh *et al.* developed transgenic tobacco by transformation with *Agrobacterium*
- 1987 - Klien *et al.* developed biolistic gene transfer method for plant transformation
- 2005 - Rice genome sequenced under International Rice Genome Sequencing Project

3. Basics of plant cell and tissue culture

In plant cell culture, plant tissues and organs are grown *in vitro* on artificial media, under aseptic and controlled environment. The technique depends mainly on the concept of totipotentiality of plant cells [9] which refers to the ability of a single cell to express the full genome by cell division. Along with the totipotent potential of plant cell, the capacity of cells to alter their metabolism, growth and development is also equally important and crucial to regenerate the entire plant [1]. Plant tissue culture medium contains all the nutrients required for the normal growth and development of plants. It is mainly composed of macronutrients, micronutrients, vitamins, other organic components, plant growth regulators, carbon source and some gelling agents in case of solid medium [10]. Murashige and Skoog medium (MS medium) is most extensively used for the vegetative propagation of many plant species *in vitro*. The pH of the media is also important that affects both the growth of plants and activity of plant growth regulators. It is adjusted to the value between 5.4 - 5.8. Both the solid and liquid medium can be used for culturing. The composition of the medium, particularly the plant hormones and the nitrogen source has profound effects on the response of the initial explant.

Plant growth regulators (PGR's) play an essential role in determining the development pathway of plant cells and tissues in culture medium. The auxins, cytokinins and gibberellins are most commonly used plant growth regulators. The type and the concentration of hormones used depend mainly on the species of the plant, the tissue or organ cultured and the objective of the experiment [11]. Auxins and cytokinins are most widely used plant growth regulators in plant tissue culture and their amount determined the type of culture established or regenerated. The high concentration of auxins generally favors root formation, whereas the high concentration of cytokinins promotes shoot regeneration. A balance of both auxin and cytokinin leads to the development of mass of undifferentiated cells known as callus.

Maximum root induction and proliferation was found in *Stevia rebaudiana*, when the medium is supplemented with 0.5 mg/l NAA [12]. Cytokinins generally promote cell division and induce shoot formation and axillary shoot proliferation. High cytokinin to auxin ratio promotes shoot proliferation while high auxin to cytokinins ratio results in root formation [13]. Shoot initiation and proliferation was found maximum, when the callus of black pepper was shifted to medium supplemented with BA at the concentration of 0.5 mg/l [14]. Gibberellins are used for enhanced growth and to promote cell elongation. Maximum shoot length was observed in *Phalaenopsis* orchids when cultured in medium containing 0.5 mg/l GA_3 (unpublished).

4. Tissue culture in agriculture

As an emerging technology, the plant tissue culture has a great impact on both agriculture and industry, through providing plants needed to meet the ever increasing world demand. It has made significant contributions to the advancement of agricultural sciences in recent times and today they constitute an indispensable tool in modern agriculture [5].

Biotechnology has been introduced into agricultural practice at a rate without precedent. Tissue culture allows the production and propagation of genetically homogeneous, disease-free plant material [37]. Cell and tissue *in vitro* culture is a useful tool for the induction of somaclonal variation [38]. Genetic variability induced by tissue culture could be used as a source of variability to obtain new stable genotypes. Interventions of biotechnological approaches for *in vitro* regeneration, mass micropropagation techniques and gene transfer studies in tree species have been encouraging. *In vitro* cultures of mature and/or immature zygotic embryos are applied to recover plants obtained from inter-generic crosses that do not produce fertile seeds [39]. Genetic engineering can make possible a number of improved crop varieties with high yield potential and resistance against pests. Genetic transformation technology relies on the technical aspects of plant tissue culture and molecular biology for:

- Production of improved crop varieties
- Production of disease-free plants (virus)
- Genetic transformation
- Production of secondary metabolites
- Production of varieties tolerant to salinity, drought and heat stresses

5. Germplasm conservation

In vitro cell and organ culture offers an alternative source for the conservation of endangered genotypes [40]. Germplasm conservation worldwide is increasingly becoming an essential activity due to the high rate of disappearance of plant species and the increased need for safeguarding the floristic patrimony of the countries [41]. Tissue culture protocols can be used for preservation of vegetative tissues when the targets for conservation are clones instead of seeds, to keep the genetic background of a crop and to avoid the loss of the conserved patrimony due to natural disasters, whether biotic or abiotic stress [42]. The plant species which do not produce seeds (sterile plants) or which have 'recalcitrant' seeds that cannot be stored for long period of time can successfully be preserved via *in vitro* techniques for the maintenance of gene banks.

Cryopreservation plays a vital role in the long-term *in vitro* conservation of essential biological material and genetic resources. It involves the storage of *in vitro* cells or tissues in liquid nitrogen that results in cryo-injury on the exposure of tissues to physical and chemical stresses. Successful cryopreservation is often ascertained by cell and tissue survival and the ability to re-grow or regenerate into complete plants or form new colonies [43]. It is desirable to assess the genetic integrity of recovered germplasm to determine whether it is 'true-to-type' following cryopreservation [44]. The fidelity of recovered plants can be assessed at phenotypic, histological, cytological, biochemical and molecular levels, although, there are advantages and limitations of the various approaches used to assess genetic stability [45]. Cryobionomics is a new approach to study genetic stability in the cryopreserved plant materials [46]. The embryonic tissues can be cryopreserved for future use or for germplasm conservation [47].

6. Embryo culture

Embryo culture is a type of plant tissue culture that is used to grow embryos from seeds and ovules in a nutrient medium. In embryo culture, the plant develops directly from the embryo or indirectly through the formation of callus and then subsequent formation of shoots and roots. The technique has been developed to break seed dormancy, test the vitality of seeds, production of rare species and haploid plants [59, 119]. It is an effective technique that is employed to shorten the breeding cycle of plants by growing excised embryos and results in the reduction of long dormancy period of seeds. Intra-varietal hybrids of an economically important energy plant "Jatropha" have been produced successfully with the specific objective of mass multiplication [62]. Somatic embryogenesis and plant regeneration has been carried out in embryo cultures of Jucara Palm for rapid cloning and improvement of selected individuals [60]. In addition, conservation of endangered species can also be attained by practicing embryo culture technique. Recently a successful protocol has been developed for the *in vitro* propagation of *Khaya grandifoliola* by excising embryos from mature seeds [61]. The plant has a high economic value for timber wood and for medicinal purposes as well. This technique has an important application in forestry by offering a mean of propagation of elite individuals where the selection and improvement of natural population is difficult.

7. Genetic transformation

Genetic transformation is the most recent aspect of plant cell and tissue culture that provides the mean of transfer of genes with desirable trait into host plants and recovery of transgenic plants [63]. The technique has a great potential of genetic improvement of various crop plants by integrating in plant biotechnology and breeding programmes. It has a promising role for the introduction of agronomically important traits such as increased yield, better quality and enhanced resistance to pests and diseases [64].

Genetic transformation in plants can be achieved by either vector-mediated (indirect gene transfer) or vector less (direct gene transfer) method [65]. Among vector dependant gene transfer methods, *Agrobacterium*-mediated genetic transformation is most widely used for the expression of foreign genes in plant cells. Successful introduction of agronomic traits in plants was achieved by using root explants for the genetic transformation [66]. Virus-based vectors offers an alternative way of stable and rapid transient protein expression in plant cells thus providing an efficient mean of recombinant protein production on large scale [67].

Recently successful transgenic plants of Jatropha were obtained by direct DNA delivery to mature seed-derived shoot apices via particle bombardment method [68]. This technology has an important impact on the reduction of toxic substances in seeds [69] thus overcoming the obstacle of seed utilization in various industrial sector. Regeneration of disease or viral resistant plants is now achieved by employing genetic transformation technique. Researchers succeeded in developing transgenic plants of potato resistant to potato virus Y (PVY) which is a major threat to potato crop worldwide [70]. In addition, marker free transgenic plants of *Petunia hybrida* were produced using multi-auto-transformation (MAT) vector system. The plants exhibited high level of resistance to *Botrytis cinerea,* causal agent of gray mold [71].

8. Protoplast fusion

Somatic hybridization is an important tool of plant breeding and crop improvement by the production of interspecific and intergeneric hybrids. The technique involves the fusion of protoplasts of two different genomes followed by the selection of desired somatic hybrid cells and regeneration of hybrid plants [48]. Protoplast fusion provides an efficient mean of gene transfer with desired trait from one species to another and has an increasing impact on crop improvement [3]. Somatic hybrids were produced by fusion of protoplasts from rice and ditch reed using electrofusion treatment for salt tolerance [49].

In vitro fusion of protoplast opens a way of developing unique hybrid plants by overcoming the barriers of sexual incompatibility. The technique has been applicable in horticultural industry to create new hybrids with increased fruit yield and better resistance to diseases. Successful viable hybrid plants were obtained when protoplasts from citrus were fused with other related citrinae species [50]. The potential of somatic hybridization in important crop plants is best illustrated by the production of intergeneric hybrid plants among the members of Brassicaceae [51]. To resolve the problem of loss of chromosomes and decreased regeneration capacity, successful protocol has been established for the production of somatic hybrid plants by using two types of wheat protoplast as recipient and protoplast of *Haynaldia villosa* as a fusion donor. It is also employed as an important gene source for wheat improvement [52].

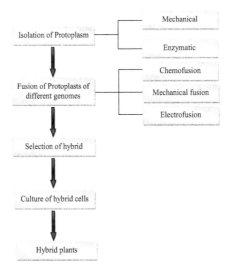

Figure 1. Schematic representation of production of hybrid plant via protoplast fusion

9. Haploid production

The tissue culture techniques enable to produce homozygous plants in relatively short time period through the protoplast, anther and microspore cultures instead of conventional breeding [53].

Haploids are sterile plants having single set of chromosomes which are converted into homozygous diploids by spontaneous or induced chromosome doubling. The doubling of chromosomes restores the fertility of plants resulting in production of double haploids with potential to become pure breeding new cultivars [54]. The term androgenesis refers to the production of haploid plants from young pollen cells without undergoing fertilization. Sudherson et al. [55] reported haploid plant production of sturt's desert pea by using pollen grains as primary explants. The haploidy technology has now become an integral part of plant breeding programs by speeding up the production of inbred lines [56] and overcoming the constraints of seed dormancy and embryo non-viability [57]. The technique has a remarkable use in genetic transformation by the production of haploid plants with induced resistance to various biotic and abiotic stresses. Introduction of genes with desired trait at haploid state followed by chromosome doubling led to the production of double haploids inbred wheat and drought tolerant plants were attained successfully [58].

10. Current and future status of plant tissue culture

The past decades of plant cell biotechnology has evolved as a new era in the field of biotechnology, focusing on the production of a large number of secondary plant products. During the second half of the last century the development of genetic engineering and molecular biology techniques allowed the appearance of improved and new agricultural products which have occupied an increasing demand in the productive systems of several countries worldwide [31, 32, 33, 34]. Nevertheless, these would have been impossible without the development of tissue culture techniques, which provided the tools for the introduction of genetic information into plant cells [35]. Nowadays, one of the most promising methods of producing proteins and other medicinal substances, such as antibodies and vaccines, is the use of transgenic plants [36]. Transgenic plants represent an economical alternative to fermentation-based production systems. Plant-made vaccines or antibodies (plantibodies) are especially striking, as plants are free of human diseases, thus reducing screening costs for viruses and bacterial toxins. The number of farmers who have incorporated transgenic plants into their production systems in 2008 was 13.3 million, in comparison to 11 million in 2007 [34].

11. Techniques of plant tissue culture

11.1. Micropropagation

Micropropagation starts with the selection of plant tissues (explant) from a healthy, vigorous mother plant [15].Any part of the plant (leaf, apical meristem, bud and root) can be used as explant. The whole process can be summarized into the following stages as shown in Figure 2.

11.2. Stage 0: Preparation of donor plant

Any plant tissue can be introduced *in vitro*. To enhance the probability of success, the mother plant should be *ex vitro* cultivated under optimal conditions to minimize contamination in the *in vitro* culture [16].

11.3. Stage I: Initiation stage

In this stage an explant is surface sterilized and transferred into nutrient medium. Generally, the combined application of bactericide and fungicide products is suggested. The selection of products depends on the type of explant to be introduced. The surface sterilization of explant in chemical solutions is an important step to remove contaminants with minimal damage to plant cells [17]. The most commonly used disinfectants are sodium hypochlorite [18, 19], calcium hypochlorite [20], ethanol [21] and mercuric chloride ($HgCl_2$) [17]. The cultures are incubated in growth chamber either under light or dark conditions according to the method of propagation.

11.4. Stage II: Multiplication stage

The aim of this phase is to increase the number of propagules [22]. The number of propagules is multiplied by repeated subcultures until the desired (or planned) number of plants is attained.

11.5. Stage III: Rooting stage

The rooting stage may occur simultaneously in the same culture media used for multiplication of the explants. However, in some cases it is necessary to change media, including nutritional modification and growth regulator composition to induce rooting and the development of strong root growth.

11.6. Stage IV: Acclimatization Stage

At this stage, the *in vitro* plants are weaned and hardened. Hardening is done gradually from high to low humidity and from low light intensity to high light intensity. The plants are then transferred to an appropriate substrate (sand, peat, compost etc.) and gradually hardened under greenhouse.

12. Somatic embryogenesis and organogenesis

Somatic embryogenesis: is an *in vitro* method of plant regeneration widely used as an important biotechnological tool for sustained clonal propagation [23]. It is a process by which somatic cells or tissues develop into differentiated embryos. These somatic embryos can develop into whole plants without undergoing the process of sexual fertilization as done by zygotic embryos. The somatic embryogenesis can be initiated directly from the explants or indirectly by the establishment of mass of unorganized cells named callus.

Plant regeneration via somatic embryogenesis occurs by the induction of embryogenic cultures from zygotic seed, leaf or stem segment and further multiplication of embryos. Mature embryos are then cultured for germination and plantlet development, and finally transferred to soil

Somatic embryogenesis has been reported in many plants including trees and ornamental plants of different families. The phenomenon has been observed in some cactus species [24]. There are various factors that affect the induction and development of somatic embryos in cultured cells. A highly efficient protocol has been reported for somatic embryogenesis on grapevine [25] that showed higher plant regeneration sufficiently when the tissues were cultured in liquid medium. Plant growth regulators play an important role in the regeneration and proliferation of somatic embryos. Highest efficiency of embryonic callus was induced by culturing nodal stem segments of rose hybrids on medium supplemented with various PGR's alone or in combination [26]. This embryonic callus showed high germination rate of somatic embryos when grown on abscisic acid

(ABA) alone. Somatic embryogenesis is not only a process of regenerating the plants for mass propagation but also regarded as a valuable tool for genetic manipulation. The process can also be used to develop the plants that are resistant to various kinds of stresses [27] and to introduce the genes by genetic transformation [28]. A successful protocol has been developed for regeneration of cotton cultivars with resistance to *Fusarium* and *Verticillium* wilts [29].

Organogenesis: refers to the production of plant organs i.e. roots, shoots and leaves that may arise directly from the meristem or indirectly from the undifferentiated cell masses (callus). Plant regeneration via organogenesis involves the callus production and differentiation of adventitious meristems into organs by altering the concentration of plant growth hormones in nutrient medium. Skoog and Muller [30] were the first who demonstrated that high ratio of cytokinin to auxin stimulated the formation of shoots in tobacco callus while high auxin to cytokinin ratio induced root regeneration.

Figure 2. Flow chart summarizing tissue culture experiments.

13. Tissue culture in pharmaceuticals

Plant cell and tissue cultures hold great promise for controlled production of myriad of useful secondary metabolites [72]. Plant cell cultures combine the merits of whole-plant systems with those of microbial and animal cell cultures for the production of valuable therapeutic secondary metabolites [73]. In the search for alternatives to production of medicinal compounds from plants, biotechnological approaches, specifically plant tissue cultures, are found to have potential as a supplement to traditional agriculture in the industrial production of bioactive plant metabolites [74]. Exploration of the biosynthetic capabilities of various cell cultures has been carried out by a group of plant scientists and microbiologists in several countries during the last decade [75].

Cell suspension culture: Cell suspension culture systems are used now days for large scale culturing of plant cells from which secondary metabolites could be extracted. A suspension culture is developed by transferring the relatively friable portion of the callus into liquid medium and is maintained under suitable conditions of aeration, agitation, light, temperature and other physical parameters [76]. Cell cultures cannot only yield defined standard phytochemicals in large volumes but also eliminate the presence of interfering compounds that occur in the field-grown plants [77]. The advantage of this method is that it can ultimately provide a continuous, reliable source of natural products [78]. The major advantage of the cell cultures include synthesis of bioactive secondary metabolites, running in controlled environment, independently from climate and soil conditions [79].A number of different types of bioreactors have been used for mass cultivation of plant cells. The first commercial application of large scale cultivation of plant cells was carried out in stirred tank reactors of 200 liter and 750 liter capacities to produce shikonin by cell culture of *Lithospermum erythrorhizon* [80]. Cell of *Catharanthus roseus, Dioscorea deltoidea, Digitalis lanata, Panax notoginseng, Taxus wallichiana* and *Podophyllum hexandrum* have been cultured in various bioreactors for the production of secondary plant products.

A number of medicinally important alkaloids, anticancer drugs, recombinant proteins and food additives are produced in various cultures of plant cell and tissues. Advances in the area of cell cultures for the production of medicinal compounds has made possible the production of a wide variety of pharmaceuticals like alkaloids, terpenoids, steroids, saponins, phenolics, flavanoids and amino acids [72, 81]. Some of these are now available commercially in the market for example shikonin and paclitaxel (Taxol). Until now 20 different recombinant proteins have been produced in plant cell culture, including antibodies, enzymes, edible vaccines, growth factors and cytokines [73]. Advances in scale-up approaches and immobilization techniques contribute to a considerable increase in the number of applications of plant cell cultures for the production of compounds with a high added value. Some of the secondary plant products obtained from cell suspension culture of various plants are given in Table 1.

Secondary metabolite	Plant name	Reference
Vasine	*Adhatoda vasica*	[82]
Artemisinin	*Artemisia annua*	[83]
Azadirachtin	*Azadirachta indica*	[84]
Cathin	*Brucea javanica*	[85]
Capsiacin	*Capsicum annum*	[86]
Sennosides	*Cassia senna*	[87]
Ajmalicine Secologanin Indole alkaloids Vincristine	*Catharanthus roseus*	[88] [89] [90] [91]
Stilbenes	*Cayratia trifoliata*	[92]
Berberin	*Coscinium fenustratum*	[93]
Sterols	*Hyssopus officinalis*	[94]
Shikonin	*Lithospermum erythrorhizon*	[95]
Ginseng saponin	*Panax notoginseng*	[96]
Podophyllotoxin	*Podophyllum hexandrum*	[97]
Taxane Paclitaxel	*Taxus chinensis*	[98]

Table 1. List of some secondary plant product produced in suspension culture

14. Hairy root cultures

The hairy root system based on inoculation with Agrobacterium rhizogenes has become popular in the last two decades as a method of producing secondary metabolites synthesized in plant roots [99]. Organized cultures, and especially root cultures, can make a significant contribution in the production of secondary metabolites. Most of the research efforts that use differentiated cultures instead of cell suspension cultures have focused on transformed (hairy) roots. Agrobacterium rhizogenes causes hairy root disease in plants. The neoplastic (cancerous) roots produced by A. rhizogenes infection are characterized by high growth rate, genetic stability and growth in hormone free media [100]. High stability [101] and productivity features allow the exploitation of hairy roots as valuable biotechnological tool for the production of plant secondary metabolites [102]. These genetically transformed root cultures can produce levels of secondary metabolites comparable to that of intact plants [103]. Hairy root technology has been strongly improved by increased knowledge of molecular mechanisms underlying their development. Optimizing the composition of nutrients for hairy root cultures is critical to gain a high production of secondary metabolites [100]. Some of the secondary plant products obtained from hairy root culture of various plants are shown in Table 2.

Secondary metabolite	Plant name	Reference
Rosmarinic acid	*Agastache rugosa*	[104]
Deoursin	*Angelica gigas*	[105]
Resveratol	*Arachys hypogaea*	[106]
Tropane	*Brugmansia candida*	[107]
Asiaticoside	*Centella asiatica*	[108]
Rutin	*Fagopyrum esculentum*	[104]
Glucoside	*Gentiana macrophylla*	[109]
Glycyrrhizin	*Glycyrrhiza glabra*	[110]
Shikonin	*Lithospermum erythrorhizon*	[111]
Glycoside	*Panax ginseng*	[112]
Plumbagin	*Plumbago zeylanica*	[113]
Anthraquinone	*Rubia akane*	[114]
Silymarin	*Silybium marianum*	[115]
Flavonolignan	*Silybium mariyanm*	[116]
Vincamine	*Vinca major*	[117]
Withanoloid A	*Withania somnifera*	[118]

Table 2. List of some secondary plant product produced in Hairy root culture

15. Tissue culture facilities at Qarshi industries

Plant tissue culture Lab was established in 2004 with the objectives to raise endangered medicinal plant species and the plants difficult to raise through traditional methods for conservation and mass propagation. We have so far propagated 12 medicinal plant species (*Plumbago zeylanica* L., *Nicotiana tabacum* L., *Artemisia absinthium* L., *Rosa damascena* Mill., *Althea rosea* L.,*Stevia rebaudiana* Bertoni., *Jatropha curcas* L., *Phalaenopsis,Piper nigrum* L., *Solanum tuberosum* L., *Araucaria heterophylla* Salisb. Franco., *Taxus wallichiana* Zucc.) and currently working on propagation of commercially important endangered woody plant species like *Taxus wallichiana*. Commercialization of some fruit and vegetable crops are underway. The protocols developed for The Moth Orchid, Tobacco, Honey Plant, Potato and Physic nut are presented as case studies.

16. Case study 1

16.1. Micropropagation of Phalaenopsis "The Moth Orchids"

Orchids are usually grown for the beauty, exoticism and fragrance of their flowers. They are cultivated since the times of Confucius (ca. 551 - 479 BC). Some orchids are commercialized not for their beauty, but for uses in food industry. They are also used medicinally as a treatment for diarrhea and as an aphrodisiac. The vegetative propagation of phalaenopsis is difficult and time consuming. In addition, the desired characteristics of seedlings and uniformity are not attained.

In vitro propagation studies of phalaenopsis "the moth orchids" had the objective to develop a protocol for plant regeneration from callus. Thus *in vitro* culture techniques are adopted for quick propagation of commercially important orchid species. Regeneration from callus gives a way to rectify the problem of explants shortage. The callus of phalaenopsis previously obtained from the mature orchid plant was used as explant source. The callus was maintained on MS medium added with 3.0 % sucrose, 0.8 % agar, and different concentrations of BAP and 2, 4-D. Callus was sub-cultured after every 30 days for proliferation. Maximum callus proliferation was obtained when the medium was supplemented with 0.5 mg/l BA. Fresh green and non friable callus was obtained. For shoot regeneration and elongation, the callus was transferred to MS medium supplemented with BAP and GA_3 at different concentrations. Maximum shoot elongation was obtained in medium supplemented with 1.0 mg/l GA_3 as shown in Figure 3 a, b, c.

The regenerated shoots showed excess root development when transferred to medium added with 2.0 mg/l IBA. Further research work will focus on different potting medium compositions best suited for acclimatization of regenerated plants. As a high value crop, the mass production of orchids will provide a good opportunity of marketing locally as a good source of income.

Figure 3. Micropropagation of Orchids (a) callus culture (b) shoot regeneration (c) rooted plantlets

17. Case study 2

17.1. Tissue culture of Tobacco (*Nicotiana tabacum* L.)

Tobacco is an important crop of Pakistan which covers a large area under cultivation. Being a cash crop grown all over the world, it has a good economic value. Fresh leaves of the plants are processed to obtain an agricultural product that is commercially available in dried, cured and natural forms. Clonal propagation of four important low nicotine content hybrid varieties of tobacco i.e. PGH-01, PGH-02, PGH-04 and PGH-09 was carried out with the special objective of commercialization of tissue cultured plants to the farmers and industry. The mother plants were provided by Pakistan Tobacco Board (PTB). Leaves and meristems were used as explants for the initiation of callus culture. Callus induction and proliferation was carried out on MS medium supplemented with different concentrations of

2,4-D. Excellent growth of callus was obtained at medium containing 1.0 mg/l 2,4-D. Callus was transferred to next medium for shoot regeneration. Efficient numbers of shoots were obtained when culture was shifted to MS medium supplemented with 0.5 mg/l BAP. For root induction different concentrations of IBA and NAA were tested and the result was found best on the same medium supplemented with 2.0 mg/l IBA as shown in Figure 4 a, b, c.

Figure 4. Tissue culture of *Nicotiana tabacum* (a) callus (b) shoot regeneration (c) root induction

18. Case study 3

In vitro **propagation of Honey Plant (*Stevia rebaudiana* Bertoni)** The *in vitro* clonal propagation of *Stevia rebaudiana* was conducted by inoculating seeds on MS medium [10] and placing under photoperiod of 16 hrs light and 8 hrs dark in growth room. The seedlings with four nodes have been divided into 0.5 cm pieces of nodal segments and used as explants. For shoot multiplication, the nodal explants were inoculated on MS medium supplemented with 3.0% sucrose and 0.5, 1.0, 2.0, 3.0 and 4.0 mg/l of BAP and Kn (Kinetin) alone or in combinations with 0.25 and 0.5 mg/l of IAA. MS medium containing 2.0 mg/l BAP showed the best response to multiple shoot formation, while the highest shoot length (3.73 ± 0.14 cm) per micro shoot was observed on MS medium containing 2.0 Kn and 0.25 mg/l IAA after 15 days of inoculation as shown in Figure 5 a, b, c. Excised micro shoots were cultured on MS medium supplemented with 0.25, 0.5, 1.0 and 1.5 mg/l of NAA and IBA separately for the root induction. The optimal rooting (81%) was observed on MS medium containing 0.5 mg/l NAA with 2 % sucrose within two weeks of culture transfer. The rooted plantlets were acclimatized successfully and transferred to greenhouse under low light intensity. This protocol for *in vitro* clonal propagation of *Stevia rebaudiana* has been optimized for the local environment, as a consequence it will be helpful to establish and cultivate *Stevia rebaudiana* for commercial scale production in various environmental conditions in Pakistan.

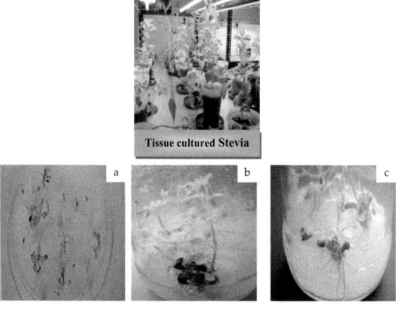

Figure 5. *In vitro* propagation of *S. rebaudiana* (a) seed germination on MS medium (b) shoot multiplication (c) root development.

19. Case study 4

19.1. Multiplication and regeneration of Potato (*Solanum tuberosum* L.) from nodal explants

Solanum tuberosum L. (potato) is the most important vegetable crop that occupies major area under cultivation in Pakistan. The crop is high yielding, has high nutritive value and gives maximum returns to farmers. Tissue culture is employed as a technique for rapid multiplication of potato plants free from diseases. The research was carried out with the objective of mass multiplication of true-to type three potato varieties i.e. Desiree, Diamant and Cardinal. The plant material for this research was provided by Four Brothers Agri Services Pakistan. The Company is working for introduction of high yielding vegetable & crop varieties in Pakistan.

The disease free potato tubers were washed both with detergent and distilled water to remove impurities and allowed to sprouting. Five days old sprouts were used as explants for direct proliferation. The explants were surface sterilized in detergent for 10 minutes, later with 0.1 % mercuric chloride solution for 5 minutes followed by three times washing with sterilized-distilled water. The sprouts were aseptically cut into 10 mm sections containing one node and inoculated in medium. The Espinosa medium plus vitamin B5 supplemented with different concentrations of BAP and GA3 alone and in combinations was utilized. Highest shoot length of shoots was observed in presence of 0.5 mg/l BAP and 0.4 mg/l GA3 with the ability to produce maximum plantlets per explant. For root induction the same medium was used with different concentrations of NAA and IBA. NAA at 2.0 mg/l induced the highest root development. The rooted plantlets were successfully acclimatized and delivered to the company for cultivation.

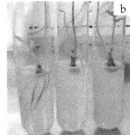

Figure 6. Tissue culture of Potato (a) nodal segment (b) regenerated shoots and roots (c) tissue cultured potato

20. Case study 5

20.1. Tissue culture of physic nut (*Jatropha curcas* L.)

The research studies on Tissue Culture of Jatropha (physic nut) had the objectives to develop protocol for mass propagation of elite trees selected on the bases of higher seed production and oil content. The experimental plant of *Jatropha curcas* was grown in the laboratory under controlled conditions for *in vitro* studies.

Leaf and apical meristem explants isolated from 7 days old seedling of *Jatropha curcas*, were use to induce callus.

Murashige & Skoog (1962) medium supplemented with different growth regulator formulations including 2,4-D and IBA was used. Excellent growth of callus on leaf explants was obtained in medium supplemented with 1.0 mg/L 2, 4-D. Callus produced from leaf explants in all IBA concentrations grew faster during 7 to 30 days of culture and then stabilized at a slow growth rate. While 1.0 mg/L 2,4-D was proved to be most effective in inducing callus on a large scale in short period of time. Callus was soft, friable and white in color.

Apical meristem was used as explant for direct shoot regeneration. Rooting from meristem was effectively achieved on MS supplemented with 1.5, 2.0 and 2.5 mg/l IBA. Root induction with 2.0 mg/l IBA was most effective and the roots also developed secondary roots.

In near future somatic embryogenesis and shoot regeneration from callus will be tested in MS medium supplemented with various concentrations of BA. The regenerated plant will be acclimatized and released for field planting under various climatic and soil conditions for further studies.

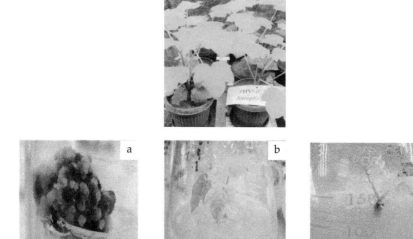

Figure 7. Tissue culture of *Jatropha curcas* (a) callus of Jatropha (b) shoot regeneration (c) root induction.

21. Conclusion

Plant tissue culture represents the most promising areas of application at present time and giving an out look into the future. The areas ranges from micropropagation of ornamental and forest trees, production of pharmaceutically interesting compounds, and plant breeding for improved nutritional value of staple crop plants, including trees to cryopreservation of valuable germplasm. All biotechnological approaches like genetic engineering, haploid induction, or somaclonal variation to improve traits strongly depend on an efficient *in-vitro* plant regeneration system.

The rapid production of high quality, disease free and uniform planting stock is only possible through micropropagation. New opportunities has been created for producers, farmers and nursery owners for high quality planting materials of fruits, ornamentals, forest tree species and vegetables. Plant production can be carried out throughout the year irrespective of season and weather. However micropropagation technology is expensive as compared to conventional methods of propagation by means of seed, cuttings and grafting etc. Therefore it is essential to adopt measures to reduce cost of production. Low cost production of plants requires cost effective practices and optimal use of equipment to reduce the unit cost of plant production. It can be achieved by improving the process efficiency and better utilization of resources. Bioreactor based plant propagation can increase the speed of multiplication and growth of cultures and reduce space, energy and labor requirements when commencing commercial propagation. However, the use of bioreactors needs special care and handling to avoid contamination of culture which may lead to heavy economic losses. The cost of production may also be reduced by selecting several plants that provide the option for around the year production and allow cost flow and optimal use of equipment and resources. It is also essential to have sufficient mother culture and reduce the number of subculture to avoid variation and plan the production of plants according to the demand.

Quality control is also very essential to assure high quality plant production and to obtain confidence of the consumers. The selection of explants source, diseases free material, authenticity of variety and elimination of somaclonal variants are some of the most critical parameters for ensuring the quality of the plants.

The *in vitro* culture has a unique role in sustainable and competitive agriculture and forestry and has been successfully applied in plant breeding for rapid introduction of improved plants. Plant tissue culture has become an integral part of plant breeding. It can also be used for the production of plants as a source of edible vaccines. There are many useful plant-derived substances which can be produced in tissue cultures.

Since last two decades there have been considerable efforts made in the use of plant cell cultures in bioproduction, bioconversion or biotransformation and biosynthetic studies. The potential commercial production of pharmaceuticals by cell culture techniques depends upon detailed investigations into the biosynthetic sequence. There is great potential of cell culture to be use in the production of valuable secondary products. Plant tissue culture is a noble approach to obtain these substances in large scale.

Plant cell culture has made great advances. Perhaps the most significant role that plant cell culture has to play in the future will be in its association with transgenic plants. The ability to accelerate the conventional multiplication rate can be of great benefit to many countries where a disease or some climatic disaster wipes out crops. The loss of genetic resources is a common story when germplasm is held in field genebanks. Slow growth *in vitro* storage and cryopreservation are being proposed as solutions to the problems inherent in field genebanks. If possible, they can be used with field genebanks, thus providing a secure duplicate collection. They are the means by which future generations will be able to have access to genetic resources for simple conventional breeding programmes, or for the more complex genetic transformation work. As such, it has a great role to play in agricultural development and productivity.

13. Commonly used terms in tissue culture

Adventitious: development of organs such as buds, leaves, roots, shoots and somatic embryos from shoot and root tissues and callus.

Agar: Natural gelling agent made from algae

Aseptic technique: procedures used to prevent the introduction of microorganisms such as fungi, bacteria, viruses and phytoplasmas into cell, tissue and organ cultures, and cross contamination of cultures.

Autoclave: A machine capable of sterilizing by steam under pressure

Axenic culture: a culture without foreign or undesired life forms but may include the deliberate co-culture with different types of cells, tissues or organisms.

Callus: an unorganized mass of differentiated plant cells.

Cell culture: culture of cells or their maintenance *in vitro* including the culture of single cells.

Chemically defined medium: a nutritive solution or substrate for culturing cells in which each component is specified.

Clonal propagation: asexual multiplication of plants from a single individual or explant.

Clones: a group of plants propagated from vegetative parts, which have been derived by repeated propagation from a single individual. Clones are considered to be genetically uniform.

Contamination: infected by unwanted microorganisms in controlled environment

Cryopreservation: ultra-low temperature storage of cells, tissues, embryos and seeds.

Culture: A plant growing *in vitro* in a sterile environment

Differentiated: cultured cells that maintain all or much of the specialized structure and function typical of the cell type *in vivo*.

Embryo culture: *In vitro* culture of isolated mature or immature embryos.

Explant: an excised piece or part of a plant used to initiate a tissue culture.

Ex vitro: Organisms removed from tissue culture and transplanted; generally plants to soil or potting mixture.

Hormone: Generally naturally occurring chemicals that strongly affect plant growth

In Vitro: To be grown in glass

In Vivo: To be grown naturally

Laminar Flow Hood: An enclosed work area where the air is cleaned using HEPA filters

Medium: a solid or liquid nutritive solution used for culturing cells

Meristem: a group of undifferentiated cells situated at the tips of shoots, buds and roots, which divide actively and give rise to tissue and organs.

Micropropagation: multiplication of plants from vegetative parts by using tissue culture nutrient medium.

Propagule: a portion of an organism (shoot, leaf, callus, etc.) used for propagation.

Somatic embryos: non-zygotic bipolar embryo-like structures obtained from somatic cells.

Subculture: the aseptic division and transfer of a culture or portion of that culture to a fresh synthetic media.

Tissue culture: *in vitro* culture of cells, tissues, organs and plants under aseptic conditions on synthetic media.

Totipotency: capacity of plant cells to regenerate whole plants when cultured on appropriate media.

Transgenic: plants that have a piece of foreign DNA

Undifferentiated: cells that have not transformed into specialized tissues

14. Abbreviations

BAP	6-benylaminopurine
2,4-D	2,4-dichlorophenoxyacetic acid
EDTA	Ethylenediaminetetraacetic acid
EtOH	Ethanol
GA₃	Gibberellic acid
IAA	Indole-3-acetic acid
IBA	Indole-3-butyric acid
NAA	Naphthaleneacetic acid
KN	Kinetin

Author details

Altaf Hussain*
Qarshi University, Lahore, Pakistan

Iqbal Ahmed Qarshi
Qarshi Industries (Pvt.) Ltd. Lahore, Pakistan

Hummera Nazir and Ikram Ullah
Plant tissue Culture Lab, Qarshi Herb Research Center, Qarshi Industries (Pvt.) Ltd. Hattar, Distt. Haripur, KPK, Pakistan

*Corresponding Author

15. References

[1] Thorpe T (2007) History of plant tissue culture. J. Mol. Microbial Biotechnol. 37: 169-180.

[2] Akin-Idowu PE, Ibitoye DO, Ademoyegun OT (2009) Tissue culture as a plant production technique for horticultural crops. Afr. J. Biotechnol. 8(16): 3782-3788.

[3] Brown DCW, Thorpe TA (1995) Crop improvement through tissue culture. World J. Microbiol & Biotechnol. 11: 409-415.

[4] George EF (1993) Plant propagation by Tissue Culture. Eastern Press, Eversley.

[5] Garcia-Gonzales R, Quiroz K, Carrasco B, Caligari P (2010) Plant tissue culture: Current status, opportunities and challenges. Cien. Inv. Agr. 37(3): 5-30.

[6] Sagare AP, Lee YL, Lin TC, Chen CC, Tsay HS (2000) Cytokinin-induced somatic embryogenesis and plant regeneration in *Coryodalis yanhusuo* (Fumariaceae)- a medicinal plant. Plant Sci. 160: 139-147.

[7] El-Dougdoug KA, El-Shamy MM (2011) Management of viral diseases in banana using certified and virus tested plant material. Afr. J. Microbiol. Res. 5(32): 5923-5932.

[8] Singh RB (1992) Current status and future prospects of plant biotechnologies in developing countries in Asia. In: Sasson A, Costarini, editors. Plant Biotechnologies for Developing Countries. London: 141-162.

[9] Haberlandt G (1902) Kulturversuche mit isolierten Pflanzenzellen. Sitzungsber. Akad. Wiss. Wien. Math.-Naturwiss. Kl. Abt. J. 111: 69-92.

[10] Murashige T, Skoog F (1962) A revised medium for rapid growth and bioassays with tobacco tissue cultures. Plant Physiol. 15: 473-497.

[11] Ting IP (1982) Plant physiology. Addison-Wesleyn Reading, Massachusetts. 642.

[12] Rafiq M, Dahot MU, Mangrio SM, Naqvi HA Qarshi IA (2007) *In vitro* clonal propagation and biochemical analysis of field established *Stevia rebaudiana* Bertoni. Pak. J. Bot. 39(7): 2467-2474.

[13] Rout GR. (2004) Effect of cytokinins and auxin on micropropagation of *Clitoria ternatea* L. Biol. Lett. 41(1): 21-26.

[14] Hussain A, Naz S, Nazir H, Shinwari ZK (2011) Tissue culture of black pepper (*Piper nigrum* L.) in Pakistan. Pak. J. Bot. 43(2): 1069-1078.

[15] Murashige T (1974) Plant propagation through tissue culture. Ann. Rev. Plant Physiol. 25: 135.

[16] Cassells AC, Doyle BM (2005) Pathogen and biological contamination management: the road ahead. In: Loyola-Vargas VM, Vázquez-Flota F, editors. Plant Cell Culture Protocols, Humana Press. New York, USA: 35-50.

[17] Husain MK, Anis M (2009) Rapid *in vitro* multiplication of *Melia azedarach* L. (a multipurpose woody tree). Acta Physiologiae Plantarum. 31(4): 765-772.

[18] Tilkat E, Onay A, Yildirim H, Ayaz E (2009) Direct plant regneration from mature leaf explants of pistachio, *Pistacia vera* L. Scientia Hort. 121(3): 361-365.

[19] Marana JP, Miglioranza E, De Faria RT (2009) *In vitro* establishment of *Jacaratia spinosa* (Aubl.) ADC. Semina-Ciencias Agrarias. 30(2): 271-274.

[20] Garcia R, Morán R, Somonte D, Zaldúa Z, López A, Mena CJ (1999) Sweet potato (*Ipomoea batatas* L.) biotechnology: perspectives and progress. In: Altman A, Ziv M,

Shamay I, editors. Plant biotechnology and *in vitro* biology in 21st century. The Netherlands: 143-146.

[21] Singh KK, Gurung B (2009) *In vitro* propagation of R. maddeni Hook. F. an endangered Rhododendron species of Sikkim Himalaya. Notulae Botanicae Horti Agrobotanici Cluj-Napoca. 37(1): 79-83.

[22] Saini R, Jaiwal PK (2002) Age, position in mother seedling, orientation, and polarity of the epicotyl segments of blackgram (*Vigna mungo* L. Hepper) determines its morphogenic response. Plant Sci. 163(1): 101-109.

[23] Park YS, Barrett JD, Bonga JM (1998) Application of somatic embryogenesis in high value clonal forestry: development, genetic control and stability of cryopreserved clones. In vitro Cell. Dev. Biol. Plant. 34: 231-239.

[24] Torres-Munoz L, Rodriguez-Garay B (1996) Somatic embryogenesis in the threatened cactus *Turbinicarpus psudomacrochele* (Buxbaum & Backerberg). J. PACD. 1: 36-38.

[25] Jayasankar S, Gray DJ, Litz RE (1999) High-efficiency somatic embryogenesis and plant regeneration from suspension cultures of grapevine. Plant Cell Rep. 18: 533-537.

[26] Xiangqian LI, Krasnyanski FS, Schuyler SK (2002) Somatic embryogenesis, secondary somatic embryogenesis and shoot organogenesis in Rosa. Plant Physiol. 159: 313-319.

[27] Bouquet A, Terregrosa L (2003) Micropropagation of grapevine (*Vitus spp.*). In: Jain SM, Ishii K, editors. Micropropagation of woody trees and fruits. The Netherlands. 75: 319-352.

[28] Maynard C, Xiang Z, Bickel S, Powell W (1998) Using genetic engineering to help save American chestnut: a progress report. J. Am Chestnut Found. 12: 40-56.

[29] Han GY, Wang XF, Zhang GY, Ma ZY (2009) Somatic embryogenesis and plant regeneration of recalcitrant cottons (*Gossypium hirsutum*). Afr. J. Biotechnol. 8(3): 432-437.

[30] Skoog F, Miller CO (1957) Chemical regulation of growth and organ formation in plant tissue cultures *in vitro*. Symp. Soc. Exp. Biol. 11: 118-131.

[31] Vasil IK (1994) Molecular Improvement of Cereals. Plant Mol. Biol. 25: 925-937.

[32] Christou P, Capell T, Kohli A, Gatehouse JA, Gatehouse AMR (2006) Recent developments and future prospects in insect pest control in transgenic crops. Trends Plant Sci. 11: 302-308.

[33] Navarro-Mastache LC (2007) Large scale commercial micropropagation in Mexico. The experience of Agromod, S.A. de C.V. Acta Horti. 748: 91-94.

[34] James C (2008) Global Status of Commercialized Biotech/ GM Crops. ISAAA Brief No. 39. Ithaca, NY. 243.

[35] Pareek LK (2005) Trends in Plant Tissue Culture and Biotechnology. Jodhpur, India. Agrobios. 350.

[36] Ferrante E, Simpson D (2001) A Review of the Progression of Transgenic Plants Used to Produce Plantibodies For Human Usage. Bio. & Biomed. Sci. Issue 1.

[37] Chatenet M, Delage C, Ripolles M, Irey M, Lockhart BLE, Rott P (2001) Detection of sugarcane yellow leaf virus in quarantine and production of virus-free sugarcane by apical meristem culture. Plant Disease. 85(11): 1177-1180.

[38] Marino G, Battistini S (1990) Leaf-callus growth, shoot regeneration and somaclonal variation in *Actinidia deliciosa*: effect of media pH. Acta Horticulturae. 280: 37-44.

[39] Ahmadi A, Azadfar D, Mofidabadi AJ (2010) Study of inter-generic hybridization possibility between *Salix aegyptica* and *Populus caspica* to achieve new hybrids. Int. J. Plant Prod. 4(2): 143-147.

[40] Sengar RS, Chaudhary R, Tyagi SK (2010) Present status and scope of floriculture developed through different biological tools. Res J. of Agri. Sci. 1(4): 306-314.

[41] Filho AR, Dal Vesco LL, Nodari RO, Lischka RW, Müller CV, Guerra MP (2005) Tissue culture for the conservation and mass propagation of *Vriesea reitzii* Leme and Costa, abromelian threatened of extinction from the Brazilian Atlantic Forest. Biodivers. Conserv. 14(8): 1799-1808.

[42] Tyagi RK, Agrawal A, Mahalakshmi C, Hussain Z, Tyagi H (2007) Low-cost media for in vitro conservation of turmeric (*Curcuma longa* L.) and genetic stability assessment using RAPD markers. In Vitro Cell. Develop. Biol. Plant. 43: 51-58.

[43] Harding K (2004) Genetic integrity of cryopreserved plant cells: a review. Cryo. Lett. 25: 3-22.

[44] Day JG (2004) Cryopreservation: fundamentals, mechanisms of damage on freezing/thawing and application in culture collections. Nova Hedwigia. 79: 191-206.

[45] Harding K, Johnston J, Benson EE (2005) Plant and algal cell cryopreservation: issues in genetic integrity, concepts. In: Benett IJ, Bunn E, Clarke H, McComb JA, editors. Cryobionomics and current European applications. In: Contributing to a Sustainable Future. Western Australia: 112-119.

[46] Harding K (2010) Plant and algal cryopreservation: issues in genetic integrity, concepts in cryobionomics and current applications in cryobiology. Aspac J. Mol. Biol. Biotechnol.18(1): 151-154.

[47] Corredoira E, San-Jose MC, Ballester A, Vieitez AM (2004) Cryopreservation of zygotic embryo axes and somatic embryos of European chestnut. Cryo Lett. 25: 33-42.

[48] Evans DA, Bravo JE (1988) Agricultural applications of protoplast fusion. In: Marby TI, editor. Plant Biotechnol. Austin: 51-91.

[49] Mostageer A, Elshihy OM (2003) Establishment of salt tolerant somatic hybrid through protoplast fusion between rice and ditch reed. Arab. J. Biotech. 6(1): 01-12.

[50] Motomura T, Hidaka T, Akihama T, Omura M (1997) Protoplast fusion for production of hybrid plants between citrus and its related genera. J. Japan. Soc. Hort. Sci. 65: 685-692.

[51] Toriyama K, Hinata K, Kameya T (1987) Production of somatic hybrid plants, "Brassicomoricandia", through protoplast fusion between *Moricandia arvensis* and *Brassica oleracea*. Plant Sci. 48(2): 123-128.

[52] Liu ZY, Chen PD, Pei GZ, Wang YN, Qin BX. Wang SL (1988) Transfer of *Haynaldia villosa* chromosomes into *Triticum aestivum*. Proceeding of the 7th International Wheat Genetics Sumposium, Cambridge, UK. 355-361.

[53] Morrison RA, Evans DA (1998). Haploid plants from tissue culture: New plant varieties in a shortened time frame. Nat. Biotechnol. 6: 684-690.

[54] Basu SK, Datta M, Sharma M, Kumar A (2011) Haploid plant production technology in wheat and some selected higher plants. Aust. J. Crop Sci. 5(9): 1087-1093.

[55] Sudherson CS, Manuel J, Al-Sabah (2008) Haploid plant production from pollen grains of sturt's desert pea via somatic embryogenesis. Am-Euras. Sci. Res. 3(1): 44-47.

[56] Bajaj YPS (1990) *In vitro* production of haploids and their use in cell genetics and plant breeding. In: Bajaj YPS, editor. Biotechnol. Agr. Forest. Berlin. 3-44.

[57] Yeung EC, Thorpe TA, Jensen CJ (1981) *In vitro* fertilization and embryo culture in plant tissue culture: Methods and Applications in Agriculture, ed. Thorpe, T.A. Academic Press, New York. 253-271.

[58] Chauhan H, Khurana P (2011) Use of double haploid technology for development of stable drought tolerant bread wheat (*Triticum aestivum* L.) transgenics. Plant Biotechnol. J. 9(3): 408-417.

[59] Holeman DJ (2009) Simple embryo culture for plant breeders: a manual of technique for the extracyion and *in vitro* germination of mature plant embryos with emphasis on the rose. First edition. Rose Hybridizers Association. 10.

[60] Guerra MP, Handro W (1988) Somatic embryogenesis and plant regeneration in embryo cultures of *Euterpe edulis* Mart. (Palmae). Plant Cell Rep. 7: 550-552.

[61] Okere AU, Adegey A (2011) *In vitro* propagation of an endangered medicinal timber species *Khaya grandifoliola* C. Dc. Afr. J. Biotechnol. 10(17): 3335-3339.

[62] Mohan N, Nikdad S, Singh G (2011) Studies on seed germination and embryo culture of *Jatropha curcas* L. under *in vitro* conditions. Research Article, Biotechnol, Bioinf, Bioeng. 1(2): 187-194.

[63] Hinchee MAW, Corbin DR, Armstrong CL, Fry JE, Sato SS, Deboer DL, Petersen WL, Armstrong TA, Connor-Wand DY, Layton JG, Horsch RB (1994) Plant transformation in Plant Cell and Tissue Culture. In: Vasil LK, Thorpe TA, editors. Dordrecht: Kluwer Academic. 231-270.

[64] Sinclair TR, Purcell LC, and Sneller CH (2004) Crop transformation and the challenge to increase yield potential. Trend Plant Sci. 9: 70-75.

[65] Sasson A (1993) Biotechnologies in developing countries, present and future, Vol.1. Paris: United Nations Educational, Scientific and Cultural Organization.

[66] Franklin G, Lakshmi SG (2003) *Agrobacterium tumefaciens*-mediated transformation of eggplant (*Solanum melongena* L.) using root explants. Plant Cell Rep. 21: 549-554.

[67] Chung SM, Manjusha V, Tzfira T (2006) Agrobacterium is not alone: gene transfer to plants by viruses and other bacteria. Trends Plant Sci. 11(1): 1-4.

[68] Purkayastha J, Sugla, T, Paul A, Maumdar P, Basu A, Solleti SK, Mohommad A, Ahmed Z, Sahoo L (2010) Efficient *in vitro* plant regeneration from shoot apices and gene transfer by particle bombardment in *Jatropha curcas*. Biologia Plantarum. 54: 13-20.

[69] Misra M, Misra AN (1993) Genetic transformation of grass pea. In: DAE Symposium on Photosynth. Plant Molecular Biology, BRNS/DAE, Govt. of India. 246-251.

[70] Bukovinszki A, Diveki Z, Csanyi M, Palkovics L, Balazs, E (2007) Engineering resistance to PVY in different potato cultivars in a marker-free transformation system using a 'Shooter mutant' *A. tumefaciens*. Plant Cell Rep. 26(4): 459-465.

[71] Khan RS, Alam SS, Munir I, Azadi P, Nakamura I, Mii M (2011) *Botrytis cinerea* resistant marker-free *Petunia hybrida* produced using the MAT vector system. Plant Cell Tissue Organ Cult. 106: 11-20.

[72] Vijayasree N, Udayasri P, Aswani KY, Ravi BB, Phani KY, Vijay VM (2010) Advancements in the Production of Secondary Metabolites. J. Nat. Prod. 3: 112-123.

[73] Hellwig S, Drossard J, Twyman RM, Fischer R (2004) Plant cell cultures for the production of recombinant proteins. Nat. Biotechnol. 22: 1415- 1422.

[74] Ramachandra SR, Ravishankar GA (2002) Plant cell cultures: Chemical factories of secondary metabolites. Biotechnol. Adv. 20:1001-153.

[75] Siahsar B, Rahimi M, Tavassoli A, Raissi AS (2011) Application of Biotechnology in Production of Medicinal Plants. J. Agric. & Environ. Sci. 11(3): 439-444.

[76] Chattopadhyay S, Farkya S, Srivastava AK, Bisaria VS (2002) Bioprocess Considerations for Production of Secondary Metabolites by Plant Cell Suspension Cultures. Biotechnol. Bioprocess Eng. 7: 138-149.

[77] Lila KM (2005) Valuable secondary products from *in vitro* culture, Secondary Products *In Vitro*, CRC Press LLC.

[78] Rao RS, Ravishankar GA (2002) Plant tissue cultures; chemical factories of secondary metabolites. Biotechnol. Adv. 20: 101-153.

[79] Karuppusamy S (2009) A review on trends in production of secondary metabolites from higher plants by *in vitro* tissue, organ and cell cultures. J. Med. Plant Res. 3(13): 1222-1239.

[80] Payne GF, Shuler ML, Brodelius P (1987) Plant cell culture. J. In: B. K. Lydensen (ed). Large Scale Cell Culture Technology. Hanser Publishers, New York, USA. 193-229.

[81] Yesil-Celiktas O, Gurel A, Vardar-Sukan F (2010) Large scale cultivation of plant cell and tissue culture in bioreactors. Transworld Res. Network. 1- 54.

[82] Shalaka DK, Sandhya P (2009) Micropropagation and organogenesis in *Adhatoda vasica* for the estimation of vasine. Pharm. Mag. 5: 539-363.

[83] Baldi A, Dixit VK (2008) Enhanced artemisinin production by cell cultures of *Artemisia annua*. Curr. Terends in Biotechnol. Pharmacol. 2: 341-348.

[84] Sujanya S, Poornasri DB, Sai I (2008) In vitro production of azadirachtin from cell suspension cultures of *Azadirachta indica*. Biosci. J. 33: 113-120.

[85] Wagiah ME, Alam G, Wiryowidagdo S, Attia K (2008) Imporved production of the indole alkaloid cathin-6-one from cell suspension cultures of *Brucea javanica* (L.) Merr. Sci. Technol. J. 1: 1-6.

[86] Umamaheswai A, Lalitha V (2007) In vitro effect of various growth hormones in *Capsicum annum* L. on the callus induction and production of Capsiacin. Plant Sci. J. 2: 545-551.

[87] Shrivastava N, Patel T, Srivastava A (2006) Biosynthetic potential of in vitro grown callus cells of *Cassia senna* L. var. senna. Curr. Sci. J. 90: 1472-1473.

[88] Zhao J, Zhu W, Hu Q (2001) Enhanced catharanthine production in *Catharanthus roseus* cell cultures by combined elicitor treatment in shake flasks and bioreactors. Enzyme Microbiol. Technol. J. 28: 673-681.

[89] Contin A, Van der Heijden R, Verpoorte R (1999) Effects of alkaloid precursor feeding and elicitation on the accumulation of secologanin in a *Catharanthus roseus* cell suspension culture. Plant Cell Tiss. Org. Cult. 56: 111-119.

[90] Moreno PRH, Van der Heijden R, Verpoorte R (1993) Effect of terpenoid precursor feeding and elicitation on formation of indole alkaloids in cell suspension cultures of *Catharanthus roseus*. Plant Cell Rep. J. 12: 702- 705.

[91] Lee-Parsons CWT, Rogce AJ (2006) Precursor limitations in methyl jasmonate-induced *Catharanthus roseus* cell cultures. Plant Cell Rep. 25: 607-612.

[92] Roat C, Ramawat KG (2009) Elicitor induced accumulation of stilbenes in cell suspension cultures of Cayratia trifoliata (L.) Domin. Plant Biotechnol. Rep. J. 3: 135-138.

[93] Khan T, Krupadanam D, Anwar Y (2008) The role of phytohormone on the production of berberine in the calli culture of an endangered medicinal plant, turmeric (*Coscinium fenustratum* L.) Afr. J. Biotechnol. 7: 3244-3246.

[94] Skrzypek Z, Wysokinsku H (2003) Sterols and titerpenes in cell cultures of *Hyssopus officinalis* L. Ver Lag der Zeitschrift fur Naturforschung. D. 312.

[95] Tabata M, Fujita Y (1985) Production of shikonin by plant cell cultures. In: M. Zaitlin, P. Day and A. Hollaender (eds). Biotechnology in Plant Science. Relevance to Agriculture in the Eighties, Acedemic Press, San Diego, USA. 207-218.

[96] Zhong JJ, Chen F, Hu WW (1999) High density cultivation of *Panax notoginseng* cells in stirred tank bioreactors for the production of ginseng biomass and ginseng saponin. Process Biochem. J. 35: 491-496.

[97] Chattopadhyay S, Farkya S, Srivastava AK, Bisaria VS (2002) Bioprocess Considerations for Production of Secondary Metabolites by Plant Cell Suspension Cultures. Biotechnol. Bioprocess Eng. 7: 138-149.

[98] Wang HQ, Yu JT, Zhong JJ (1999) Significant improvement of taxane production in suspension cultures of *Taxus chinensis* by sucrose feeding strategy. Process Biochem. J. 35: 479-483.

[99] Palazon J, Pinol MT, Cusido RM, Morales C, Bonfill M (1997) Application of transformed root technology to the production of bioactive metabolites. Recent Res. Dev. Plant Physiol. 1: 125-143.

[100] Hu ZB, Du M (2006) Hairy Root and Its Application in Plant Genetic Engineering. J. Integr. Plant Biol. 48(2): 121–127.

[101] Giri A, Narasu M (2000) Transgenic Hairy Roots: Recent Trends and Application. Biotechnol. Adv. 18: 1-22.

[102] Pistelli L, Giovannini A, Ruffoni B, Bertoli A, Pistelli L (2010) Hairy Root Cultures for Secondary Metabolites Production. ISBN: 978-1-4419-7346-7.

[103] Srivastava S, Srivastava AK (2007) Hairy root culture for mass-production of high-value secondary metabolites. Crit. Rev. Biotechnol. 27(1): 29-43.

[104] Lee SY, Cho SJ, Park MH, Kim YK, Choi JI, Park SU (2007) Growth and rutin production in hairy root culture of buck weed (*Fagopyruum esculentum*). Prep. Biochem. Biotechnol. 37: 239-246.

[105] Xu H, Kim YK, Suh SY, Udin MR, Lee SY, Park SU (2008) Deoursin production from hariy root culture of *Angelica gigas*. Korea Soc. Appl. Biol. Chem. J. 51: 349-351.

[106] Kim JS, Lee SY, Park SU (2008) Resveratol production in hairy root culture of peanut, Arachys hypogaea L. transformed with differet *Agrobacterium rhizogenes* strains. Afr. J. Biotechnol. 7: 3788-3790.

[107] Marconi PL, Selten LM, Cslcena EN, Alvarez MA, Pitta-Alvarez SI (2008) Changes in growth and tropane alkaloid production in long term culture of hairy roots of *Brugmansia candida*. Elect. J. Integrative Biosci. 3: 38-44.

[108] Kim OT, Bang KH, Shin YS, Lee MJ, Jang SJ, Hyun DY, Kim YC, Senong NS, Cha SW, Hwang B (2007) Enhanced production of asiaticoside from hairy root cultures of *Centella asitica* (L.) Urban elicited by methyl jasmonate. Plant Cell Rep. 26: 1914-1949.

[109] Tiwari KK, Trivedi M, Guang ZC, Guo GQ, Zheng GC (2007) Genetic transformation of Gentiana macrophylla with *Agrobacterium rhizogenes*: growth and production of secoiridoid glucoside gentiopicroside in transformed hairy root cultures. Plant Cell Rep. J. 26: 199-210.

[110] Mehrotra S, Kukreja AK, Khanuja SPS, Mishra BN (2008) Genetic transformation studies and scale up of hairy root culture of *Glycyrrhiza glabra* in bioreactor. 11: 717-728.

[111] Fukui H, Feroj H, Ueoka T, Kyo M (1998) Formation and secretion of a new benzoquinone by hairy root cultures of *Lithospermum erythrorhizon*. Phytochem.47: 1037-1039.

[112] Jeong GA, Park DH (2007) Enhanced secondary metabolite biosynthesis by elicitation in transformed plant root system. Appl. Biochem. Biotechnol. 130: 436-446.

[113] Verma PC, Singh D, Rahman L, Gupta MM, Banerjee S (2002) *In vitro* studies in *Plumbago zeylanica* : rapid micropropagation and establishment of higher plumbagin yeilding hairy root cultures. Plant Physiol. J. 159: 547-552.

[114] Park SU, Lee SY (2009) Anthraquinone production by hairy root culture of Rubia akane Nakai : Influence of media and auxin treatment. Sci. Res. Essay. J. 4: 690-693.

[115] Rahnama H, Hasanloo T, Shams MR, Sepehrifar R (2008) Silymarin production by hairy root culture of *Silybium marianum* (L.) Gaertn. Iranian Biotechnol. J. 6:113-118.

[116] Alikaridis F, Papadakis D, Pantelia K, Kephalas, T (2000). Flavonolignan production from *Silybium marianum* transformed and untransformed root cultures. Fitoterapia 71: 379-384.

[117] Tanaka N, Takao M, Matsumoto T (2004) Vincamine production in multiple shoot culture derived from hairy roots of *Vinca major*. Plant Cell Tiss. Org. Cult. J. 41: 61-64.

[118] Murthy HN, Dijkstra C, Anthony P, White DA, Davey MR, Powers JB, Hahn EJ, Paek KY (2008) Establishemnt of *Withania somnifera* hairy root cultures for the production of Withanoloid A. Integ. Plant Biol. J. 50:915-981.

[119] Burun B, Poyrazoglu EC (2002) Embyo culture in barley (*Hordeum vulgare* L.) Turk J. Biol. 26: 175-180.

Chapter 2

Plant Tissue Culture Media

Abobkar I. M. Saad and Ahmed M. Elshahed

Additional information is available at the end of the chapter

1. Introduction

Optimal growth and morphogenesis of tissues may vary for different plants according to their nutritional requirements. Moreover, tissues from different parts of plants may also have different requirements for satisfactory growth [1]. Tissue culture media were first developed from nutrient solutions used for culturing whole plants e.g. root culture medium of White and callus culture medium of Gautheret. White's medium was based on Uspenski and Uspenska's medium for algae, Gautheret's medium was based on Knop's salt solution [2]. Basic media that are frequently used include Murashige and Skoog (MS) medium [1], Linsmaier and Skoog (LS) medium [3], Gamborg (B₅) medium [4] and Nitsch and Nitsch (NN) medium [5].

2. Media composition

Plant tissue culture media should generally contain some or all of the following components: macronutrients, micronutrients, vitamins, amino acids or nitrogen supplements, source(s) of carbon, undefined organic supplements, growth regulators and solidifying agents. According to the International Association for Plant Physiology, the elements in concentrations greater than 0.5 $mM.l^{-1}$ are defined as macroelements and those required in concentrations less than 0.5 $mM.l^{-1}$ as microelements [6]. It should be considered that the optimum concentration of each nutrient for achieving maximum growth rates varies among species.

2.1. Macronutrients

The essential elements in plant cell or tissue culture media include, besides C, H and O, macroelements: nitrogen (N), phosphorus (P), potassium (K), calcium (Ca), magnesium (Mg) and sulphur (S) for satisfactory growth and morphogenesis. Culture media should contain at least 25-60 mM of inorganic nitrogen for satisfactory plant cell growth. Potassium is required for cell growth of most plant species. Most media contain K in the form of nitrate chloride salts at concentrations ranging between 20 and 30 mM. The optimum concentrations of P, Mg, S and Ca range from 1-3 mM if other requirements for cell growth are provided [2].

2.2. Micronutrients

The essential micronutrients (minor elements) for plant cell and tissue growth include iron (Fe), manganese (Mn), zinc (Zn), boron (B), copper (Cu) and molybdenum (Mo). Iron is usually the most critical of all the micronutrients. The element is used as either citrate or tartarate salts in culture media, however, there exist some problems with these compounds for their difficulty to dissolve and precipitate after media preparation. There has been trials to solve this problem by using ethylene diaminetetraacetic acid (EDTA)-iron chelate (FeEDTA) [1]. A procedure for preparing an iron chelate solution that does not precipitate have been also developed [7]. Cobalt (Co) and iodine (I) may be added to certain media, but their requirements for cell growth has not been precisely established. Sodium (Na) and chlorine (Cl) are also used in some media, in spite of reports that they are not essential for growth. Copper and cobalt are added to culture media at concentrations of $0.1\mu M$, iron and molybdenum at $1\mu M$, iodine at $5\mu M$, zinc at 5-30 μM, manganese at 20-90 μM and boron at 25-100 μM [2].

2.3. Carbon and energy sources

In plant cell culture media, besides the sucrose, frequently used as carbon source at a concentration of 2-5%, other carbohydrates are also used. These include lactose, galactose, maltose and starch and they were reported to be less effective than either sucrose or glucose, the latter was similarly more effective than fructose considering that glucose is utilized by the cells in the beginning, followed by fructose. It was frequently demonstrated that autoclaved sucrose was better for growth than filter sterilized sucrose. Autoclaving seems to hydrolyze sucrose into more efficiently utilizable sugars such as fructose. Sucrose was reported to act as morphogenetic trigger in the formation of auxiliary buds and branching of adventitious roots [8].

It was found that supplements of sugar cane molasses, banana extract and coconut water to basal media can be a good alternative for reducing medium costs. These substrates in addition to sugars, they are sources of vitamins and inorganic ions required growth [9, 10].

2.4. Vitamins and myo-inositol

Some plants are able to synthesize the essential requirements of vitamins for their growth. Some vitamins are required for normal growth and development of plants, they are required by plants as catalysts in various metabolic processes. They may act as limiting factors for cell growth and differentiation when plant cells and tissues are grown *In vitro* [2]. The vitamins most used in the cell and tissue culture media include: thiamin (B_1), nicotinic acid and pyridoxine (B_6). Thiamin is necessarily required by all cells for growth [11]. Thiamin is used at concentrations ranging from 0.1 to 10 mg.l^{-1}. Nicotinic acid and pyridoxine, however not essential for cell growth of many species, they are often added to culture media [12]. Nicotinic acid is used at a concentration range 0.1-5 mg.l^{-1} and pyridoxine is used at 0.1-10 mg.l^{-1}. Other vitamins such as biotin, folic acid, ascorbic acid, pantothenic acid, tocopherol (vitamin E), riboflavin, p-amino-benzoic acid are used in some cell culture media however, they are not growth limiting factors. It was recommended that vitamins should be added to culture media only when the concentration of thiamin is below the desired level or when the cells are required to be grown at low population densities [14]. Although it is not a vitamin but a carbohydrate, myo-inositol is added in small quantities to stimulate cell growth of most plant species [13]. Myo-inositol is believed to play a role in cell division because of its breakdown to ascorbic acid and pectin and incorporation into phosphoinositides and phosphatidyl-inositol. It is generally used in plant cell and tissue culture media at concentrations of 50-5000 mg.l^{-1}.

2.5. Amino acids

The required amino acids for optimal growth are usually synthesized by most plants, however, the addition of certain amino acids or amino acid mixtures is particularly important for establishing cultures of cells and protoplasts. Amino acids provide plant cells with a source of nitrogen that is easily assimilated by tissues and cells faster than inorganic nitrogen sources. Amino acid mixtures such as casein hydrolysate, L-glutamine, L-asparagine and adenine are frequently used as sources of organic nitrogen in culture media. Casein hydrolysate is generally used at concentrations between 0.25-1 g.l^{-1}. Amino acids used for enhancement of cell growth in culture media included; glycine at 2 mg.l^{-1}, glutamine up to 8 mM, asparagine at 100mg.l^{-1}, L-arginine and cysteine at 10 mg.l^{-1} and L-tyrosine at 100mg.l^{-1} [2].

2.6. Undefined organic supplements

Some media were supplemented with natural substances or extracts such as protein hydrolysates, coconut milk, yeast extract, malt extract, ground banana, orange juice and tomato juice, to test their effect on growth enhancement. A wide variety of organic extracts are now commonly added to culture media. The addition of activated charcoal is sometimes added to culture media where it may have either a beneficial or deleterious effect. Growth and differentiations were stimulated in orchids [15], onions and carrots [16, 17], tomatoes [18]. On the other hand, an inhibition of cell growth was noticed on addition of activated charcoal to culture medium of soybean [17]. Explanation of the mode of action of activated charcoal was based on adsorption of inhibitory compounds from the medium, adsorption of growth regulators from the culture medium or darkening of the medium [19]. The presence of 1% activated charcoal in the medium was demonstrated to largely increase hydrolysis of sucrose during autoclaving which cause acidification of the culture medium [20].

2.7. Solidifying agents

Hardness of the culture medium greatly influences the growth of cultured tissues (Figure 1). There are a number of gelling agents such as agar, agarose and gellan gum [21].

Figure 1. Agar-solidified medium supporting plant growth.

Agar, a polysaccharide obtained from seaweeds, is of universal use as a gelling agent for preparing semi-solid and solid plant tissue culture media. Agar has several advantages over other gelling agents; mixed with water, it easily melts in a temperature range 60-100°C and solidifies at approximately 45°C and it forms a gel stable at all feasible incubation temperatures. Agar gels do not react with media constituents and are not digested by plant enzymes. It is commonly used in media at concentrations ranging between 0.8-1.0%. Pure agar preparations are of great importance especially in experiments dealing with tissue metabolism. Agar contains Ca, Mg and trace elements on comparing different agar brands [22]; Bacto, Noble and purified agar, in concern with contaminants. The author, for example reported Bacto agar to contain 0.13, 0.01, 0.19, 0.43, 2.54, 0.17% of Ca, Ba, Si, Cl, SO_4^-, N, respectively. Impurities also included 11.0, 285.0 and 5.0 mg.l[1-] for iron, magnesium and copper as contaminants, respectively. Amounts of some contaminants were higher in purified agar than in Bacto agar of which Mg that accounted for 695.0 mg.l[1-] and Cu for 20.0 mg.l[1-].

Reduction of culture media costs is continually targeted in large-scale cultures and search for cheap alternatives provided that white flower, potato starch, rice powder were as good gelling agents as agar. It was also experienced that combination of laundry starch, potato starch and semolina in a ratio of 2:1:1 reduced costs of gelling agents by more than 70% [23].

2.8. Growth regulators

Plant growth regulators are important in plant tissue culture since they play vital roles in stem elongation, tropism, and apical dominance. They are generally classified into the following groups; auxins, cytokinins, gibberellins and abscisic acid. Moreover, proportion of auxins to cytokinins determines the type and extent of organogenesis in plant cell cultures [24].

2.8.1. Auxins

The common auxins used in plant tissue culture media include: indole-3- acetic acid (IAA), indole-3- butric acide (IBA), 2,4-dichlorophenoxy-acetic acid (2,4-D) and naphthalene- acetic acid (NAA). IAA is the only natural auxin occurring in plant tissues There are other synthetic auxins used in culture media such as 4-chlorophenoxy acetic acid or p-chloro-phenoxy acetic acid (4-CPA, pCPA), 2,4,5-trichloro-phenoxy acetic acid (2,4,5 T), 3,6-dichloro-2-methoxy- benzoic acid (dicamba) and 4- amino-3,5,6-trichloro-picolinic acid (picloram) [2].

Auxins differ in their physiological activity and in the extent to which they translocate through tissue and are metabolized. Based on stem curvature assays, eight to twelve times higher activity was reported on using 2,4-D than IAA, four times higher activity of 2,4,5 T in comparison with IAA and NAA has as doubled activity as IAA [25]. In tissue cultures, auxins are usually used to stimulate callus production and cell growth, to initiate shoots and rooting, to induce somatic embryogenesis, to stimulate growth from shoot apices and shoot stem culture. The auxin NAA and 2,4-D are considered to be stable and can be stored at 4°C for several months [26]. The solutions of NAA and 2,4-D can also be stored for several months in a refrigerator or at -20°C if storage has to last for longer periods. It is best to prepare fresh IAA solutions each time during medium preparation, however IAA solutions can be stored in an amber bottle at 4°C for no longer than a week. Generally IAA and 2,4-D are dissolved in a small volume of 95% ethyl alcohol. NAA, 2,4-D and IAA can be dissolved in a small amount of 1N NaOH. Chemical structures of some of the frequently used auxins are given in Figure (2).

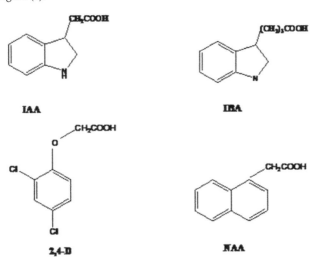

Figure 2. Chemical structure of commonly used auxins. IAA indole acetic acid, IBA Indole butyric acid, 2,4-D dichlorophenoxyacetic acid and NAA naphthalene acetic acid.

There are also some auxinlike compounds (Figure 3) that vary in their activity and are rarely used in culture media.

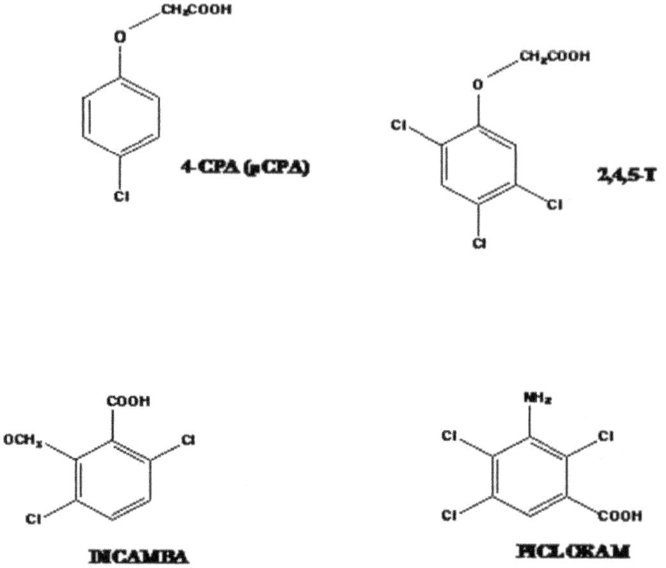

Figure 3. Some auxinlike compounds. pCPA, p-chloro-phenoxy acetic acid and 2,4,5T, 2,4,5-trichloro-phenoxy acetic acid.

2.8.2. Cytokinins

Cytokinins commonly used in culture media include BAP (6-benzyloaminopurine), 2iP (6-dimethylaminopurine), kinetin (N-2-furanylmethyl-1H-purine-6-amine), Zeatin (6-4-hydroxy-3-methyl-trans-2-butenylaminopurine) and TDZ (thiazuron-N-phenyl-N-1,2,3 thiadiazol-5ylurea). Zeatin and 2iP are naturally occurring cytokinins and zeatin is more effective. In culture media, cytokinins proved to stimulate cell division, induce shoot formation and axillary shoot proliferation and to retard root formation. The cytokinins are relatively stable compounds in culture media and can be stored desiccated at -20°C. Cytokinins are frequently reported to be difficult to dissolve and sometimes addition of few drops of 1N HCl or 1N NaOH facilitate their dissolution. Cytokinins can be dissolved in small amounts of dimethylsulfoxide (DMSO) without injury to the plant tissue [27]. DMSO has an additional advantage because it acts as a sterilizing agent; thus stock solutions containing DMSO can be added directly to the sterile culture medium. Chemical structure of the frequently used in plant tissue culture media is given in Figure (4).

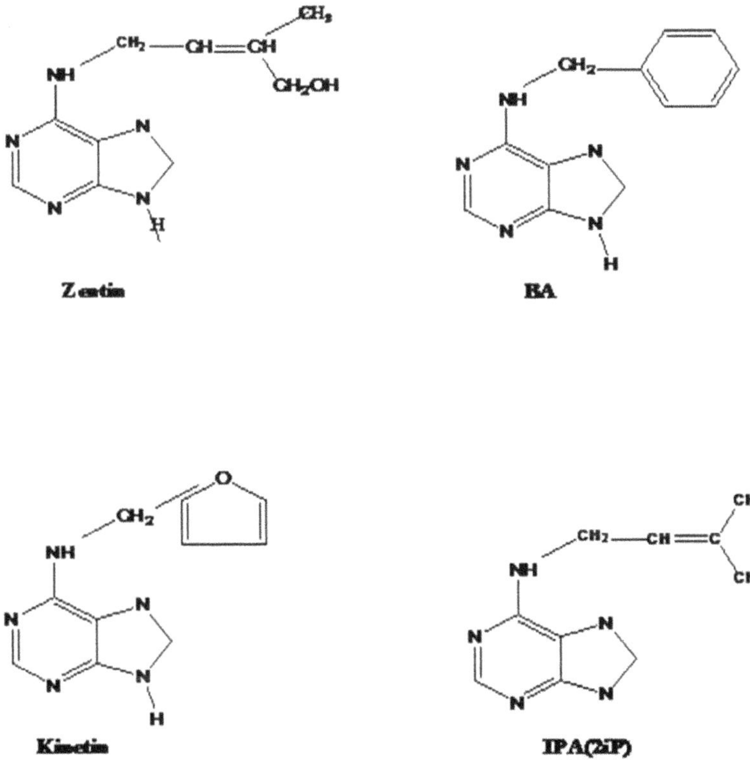

Figure 4. Chemical structure of some cytokinins, BA, benzyladenine, IPA dimethylallylamino purine.

2.8.3. Gibberellins

Gibberellins comprise more than twenty compounds, of which GA3 is the most frequently used gibberellin. These compounds enhance growth of callus [13] and help elongation of dwarf plantlets [2].

Other growth regulators are sometimes added to plant tissue culture media as abscisic acid, a compound that is usually supplemented to inhibit or stimulate callus growth, depending upon the species. It enhances shoot proliferation and inhibits later stages of embryo development [18]. Although growth regulators are the most expensive medium ingredients, they have little effect on the medium cost because they are required in very small concentrations [21].

A comparison of the chemical composition of the frequently used plant tissue culture media appears in Table (1) which was given in the appendix of the proceedings of the Technical meeting of the International Atomic Energy Agency (IAEA) [28].

Medium Components (mg.l⁻¹)	MS	G₅	W	LM	VW	Km	M	NN
Macronutrients								
Ca₃(PO4)₂					200.0			
NH₄NO₃	1650.0			400.0				720.0
KNO₃	1900.0	2500.0	80.0		525.0	180.0	180.0	950.0
CaCl₂.2H₂O	440.0	150.0		96.0				166.0
MgSO₄.7H₂O	370.0	250.0	720.0	370.0	250.0	250.0	250.0	185.0
KH₂PO₄	170.0			170.0	250.0	150.0	150.0	68.0
(NH4)₂SO4		134.0			500.0	100.0	100.0	
NaH₂PO₄.H₂O		150.0	16.5					
CaNO₃.4H₂O			300.0	556.0		200.0	200.0	
Na₂SO₄			200.0					
KCl			65.0					
K₂SO₄				990.0				
Micronutrients								
KI	0.83	0.75	0.75			80.0	0.03	
H₃BO₃	6.20	3.0	1.5	6.2		6.2	0.6	10.0
MnSO₄.4H₂O	22.30		7.0		0.75	0.075		25.0
MnSO₄.H₂O		10.0		29.43				
ZnSO₄.7H₂O	8.6	2.0	2.6	8.6			0.05	10.0
Na₂MoO₄.2H₂O	0.25	0.25		0.25		0.25	0.05	0.25
CuSO₄.5H₂O	0.025	0.025		0.25		0.025		0.025
CoCl₂.6H₂O	0.025	0.025				0.025		
Co(NO₃)₂.6H₂O							0.05	
Na₂EDTA	37.3	37.3		37.3		74.6	37.3	37.3
FeSO₄.7H₂O	27.8	27.8		27.8		25.0	27.8	27.8
MnCl₂						3.9	0.4	
Fe(C₄H₄O₆)₃.2H₂O					28.0			
Vitamins and other supplements								
Inositol	100.0	100.0		100.0				100.0
Glycine	2.0	2.0	3.0	2.0				2.0
Thiamine HCl	0.1	10.0	0.1	1.0		0.3	0.3	0.5
Pyridoxine HCl	0.5		0.1	0.5		0.3	0.3	0.5
Nicotinic acid	0.5		0.5	0.5			1.25	5.0
Ca-panthothenate			1.0					
Cysteine HCl			1.0					
Riboflavin						0.3	0.05	
Biotin							0.05	0.05
Folic acid							0.3	0.5

MS Murashige and Skoog, G₅= Gamborg *et al.*, W= White, LM= Lloyd and McCown, VW= Vacin and Went, Km= Kudson modified, M= Mitra *et al.* and NN= Nitsch and Nitsch media.

Table 1. Composition of media most frequently used.

3. Media preparation

Preparation of culture media is preferred to be performed in an equipped for this purpose compartment (Figure 5). This compartment should be constructed so as to maintain ease in cleaning and reducing possibility of contamination. Supplies of both tap and distilled water and gas should be provided. Appropriate systems for water sterilization or deionization are also important [29]. Certain devices are required for better performance such as a refrigerator, freezer, hot plate, stirrer, pH meter, electric balances with different weighing ranges, heater, Bunsen burner in addition to glassware and chemicals [30]. It is well known now that mistakes which occur in tissue culture process most frequently originate from inaccurate media preparation that is why clean glassware, high quality water, pure chemicals and careful measurement of media components should be facilitated.

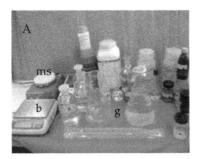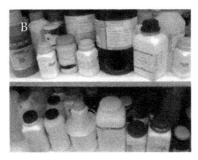

Figure 5. Some of components of the preparation room. A, some equipments used; ms magnetic stirrer hot plate, b electric balance, g glassware. B shelves for keeping chemicals.

A convenient method for preparation of culture media is to make concentrated stock solutions which can be immediately diluted to preferred concentration before use. Solutions of macronutrients are better to be prepared as stock solutions of 10 times the strength of the final operative medium. Stock solutions can be stored in a refrigerator at 2- 4ºC. Micronutrients stock solutions are made up at 100 times of the final concentration of the working medium. The micronutrients stock solution can also be stored in a refrigerator or a freezer until needed. Iron stock solution should be 100 times concentrated than the final working medium and stored in a refrigerator. Vitamins are prepared as either 100 or 1000 times concentrated stock solutions and stored in a freezer (-20ºC) until used if it is desired to keep them for long otherwise they can be stored in a refrigerator for 2-3 months and should be discarded thereafter [31]. Stock solutions of growth regulators are usually prepared at 100-1000 times the final desired concentration.

Concentrations of inorganic and organic components of media are generally expressed in mass values (mg.l^{-1}, mg/l and p.p.m.) in tissue culture literature. The International Association for plant Physiology has recommended the use of mole values. Mole is an abbreviation for gram molecular weight which is the formula weight of a substance in grams. The formula weight of a substance is equal to the sum of weights of the atoms in the chemical formula of the substance. One liter of solution containing 1mole of a substance is 1 molar (1M) or 1 mol.l^{-1} solution of the substance (1 mol.l^{-1}= 10^3 mmol.l^{-1}= 10^6 μmol/l). It is routinely now to accepted to express concentrations of macronutrients and organic nutrients in the culture medium as mmol/l values, and μmol/l values for micronutrients, hormones, vitamins and other organic constituents. This was explained on the basis that mole values for all compounds have constant number of molecules per mole [32].

4. Media selection

For the establishment of a new protocol for a specific purpose in tissue culture, a suitable medium is better formulated by testing the individual addition of a series of concentrations of a given compound to a universal basal medium such as MS, LS or B$_5$. The most effective variables in plant tissue culture media are growth regulators, especially auxins and cytokinins. Full strength of salts in media proved good for several species, but in some species the reduction of salts level to ½ or ¼ the full concentration gave better results in *in vitro* growth.

Sucrose is often assumed to be the best source of carbon for in vitro culture, the levels used are from 2 to 6% and the level has to be defined for each species.

5. Media sterilization

Prevention of contamination of tissue culture media is important for the whole process of plant propagation and helps to decrease the spread of plant parasites. Contamination of media could be controlled by adding antimicrobial agents, acidification or by filtration through microporous filters [33]. To reduce possibilities of contamination, it is recommended that sterilization rooms should have the least number of openings. Media preparation and sterilization are preferred to be performed in separate compartments. Sterilization area should also have walls and floor that withstand moisture, heat and steam [29].

Sterilization of media is routinely achieved by autoclaving at the temperature ranging from 115° – 135° C. Advantages of autoclaving are: the method is quick and simple, whereas disadvantages are the media pH changes and some components may decompose and so to loose their effectiveness. As example autoclaving mixtures of fructose, glucose and sucrose resulted in a drop in the agar gelling capacity and affecting pH of the culture medium through the formation of furfural derivatives due to sucrose hydrolysis[2].

Filtration through microporus filters (0.22- 0.45) is also used for thermolabile organic constituents such as vitamins, growth regulators and amino acids [2]. Additives of antimicrobial agents are less commonly applied in plant tissue culture media. Limitation for their use was reported and attributed to harm imposed on plants as well [34].

Author details

Abobkar I. M. Saad* and Ahmed M. Elshahed

Department of Botany and Microbiology, Faculty of Science, Sebha University, Libya

6. References

[1] Murashige T, Skoog F. A revised medium for rapid growth and bioassays with tobacco tissue cultures. Physiol. Plant. 1962; 15: 473-479.

[2] Torres KC., editor. Tissue culture techniques for horticultural crops. New York, London: Chapman and Hall; 1989.

[3] Linsmaier EM, Skoog F. Organic growth factor requirements of tobacco tissue cultures. Physiol. Plant., 1965; 18: 100-127.

[4] Gamborg OL, Miller RA, Ojima K. Nutrient requirements of suspension culture of soybean root cells. Ex. Cell. Res. 1968; 50: 15-158.

[5] Nitsch JP, Nitsch C. Haploid plants from pollen grains. Science 1969; 163: 85- 87.

[6] de Fossard R. Tissue culture for plant propagation. Armidale: University of New England; 1976.

[7] Steiner AA, Winden H. Recipe for ferric salts of ethelenediaminetetracetic acid. Plant Physiol. 1970; 46: 862- 863.

[8] Vinterhalter D, Vinterhalter BS. Micropropagation of Dracaena sp. In: Bajaj YPS (ed.) Biotechnology in agriculture and forestry 40, High-tech. and Micropropagation VI. Berlin, Heidelberg: Springer; 1997. p131- 146.

[9] Dhamankar VS. Molasses, a source of nutrients for in vitro sugar cane culture. Sugar Cane 1992; 4: 14- 15.

[10] Zahed MA M. Studies on morphogenesis of three elite species of orchids. MSc thesis, Univ. Dhaka, Bangladesh; 2000.

[11] Ohira K, Makoto I, Ojima K. Thiamine requirements of various plant cells in suspension culture. Plant Cell Physiol. 1976; 17(3): 583-590.

[12] White P R. Nutrient deficiency studies and improved inorganic nutrients for cultivation of excised tomato roots. Growth 1943; 7: 53- 65.

[13] Vasil IK, Thorpe TA, editor. Plant cell and tissue culture. Dordrecht: Kluwer Acad. Publ.; 1998.

[14] Murashige T. Plant propagation through tissue cultures. Annu. Rev. Plant Physiol. (1974); 25: 135-166.

[15] Wang WC, Yung YL, Lacis TM, Hansen JE. Greenhouse effects due to man-made perturbation of trace gases. Science 1976; 194: 685- 690.

[16] Fridborg G, Eriksson T. Effects of activated charcoal on growth and morphogenesis in cell cultures. Physiol. Plant. 1975; 34: 306- 308.

[17] Fridborg G, Pederson M, Landstrom L, Eriksson T. The effect of activated charcoal on tissue cultures: adsorption of metabolites inhibiting morphogenesis. Physiol. Plant. 1978; 43: 104- 106.

---

* Corresponding Author

[18] Anagnostakis SL. Haploid plants from anthers of tobacco enhancement with charcoal. Planta 1974; 115: 281-283.

[19] Pan MJ, van Staden J. The use of charcoal in *in vitro* culture- A review. Plant Growth Regulation 1998; 26: 155- 163.

[20] Druart Ph, Wulf O. Activated charcoal catalyses sucrose hydrolysis during autoclaving. Plant cell, tissue and organ culture 1993; 32: 97- 99.

[21] Prakash S, Hoque MI, Brinks T. Culture media and containers. In: International Atomic Energy Agency (ed.): Low cost options for tissue culture technology in developing countries. Proceedings of a technical meeting, 26-30 August 2002, Vienna, Austria.

[22] Pierik RLM. In vitro culture of higher plants. Dordrecht: Klower Acad. Publ.; 1997.

[23] Prakash S. Production of ginger and turmeric through tissue culture methods and investigations into making tissue culture propagation less expensive. PhD thesis. Bangalore Univ., India; 1993.

[24] Skoog F, Miller RA. Chemical regulations of growth and organ formation in plant tissue culture in vitro. Sym. Soc. Exp. Biol. 1957; 11: 118- 131.

[25] Lam TH, Street HE. The effect of selected aryloxalcane carboxylic acids on the growth and levels of soluble phenols in cultured cells of *Rosa damescens*. Pflanzenphysiol. 1977; 84: 121.

[26] Gamborg OL, Murashige T, Thorpe TA, Vasil IK. Plant tissue culture. In vitro cellular and developmental biology 1976; 12: 473- 478.

[27] Schmitz RY, Skoog F, Playtis AJ, Leonard NJ. Cytokinins: synthesis and biological activity of geometric and position isomers of zeatin. Plant Physiol.1972; 50: 702-705.

[28] International Atomic Energy Agency (IAEA) 2002- TECDOC-1284. Low cost options for tissue culture technology in developing countries. Proceedings of a technical meeting, August 26-30, 2002, Vienna, Austria.

[29] Ahloowali BS, Prakash J. Physical components of tissue culture technology. In: International Atomic Energy Agency (ed.): Low cost options for tissue culture technology in developing countries. Proceedings of a technical meeting, 26-30 August 2002, Vienna, Austria.

[30] Brown D, Thorpe T. Organization of a plant tissue culture laboratory. In: (ed.) Vasil I. Cell culture and somatic cell genetics of plants. Laboratory procedures and their applications. New York: Acad. Press; 1984. p 1-12

[31] [31] Gamborg OL, Shyluk JP, Shahin EA. Isolation, fusion and culture of plant protoplasts. In: (ed.) Thorpe TA. Plant Tissue Culture: Methods and Applications in Agriculture. New York: Academic Press; 1981. p115-153.

[32] Bhojwani SS, Radzan MK. Plant tissue culture: Theory and practice. Amsterdam: Elsevier; (1983):

[33] Levin R, Tanny G. Bioreactors as a low cost option for tissue culture. In: International Atomic Energy Agency (ed.): Low cost options for tissue culture technology in developing countries. Proceedings of a technical meeting, 26-30 August 2002, Vienna, Austria.

[34] Savangikar VA. Role of low cost options in tissue culture. In: International Atomic Energy Agency (ed.): Low cost options for tissue culture technology in developing countries. Proceedings of a technical meeting, 26-30 August 2002, Vienna, Austria.

Chapter 3

The Prerequisite of the Success in Plant Tissue Culture: High Frequency Shoot Regeneration

Mustafa Yildiz

Additional information is available at the end of the chapter

1. Introduction

Plant tissue culture is a term containing techniques used to propagate plants vegetatively by using small parts of living tissues (explants) on artificial growth mediums under sterile conditions. Explants regenerate shoots and roots, and consequently whole fertile plants under certain cultural conditions. Micropropagation is the production of whole plants through tissue culture from small parts such as shoot and root tips, leaf tissues, anthers, nodes, meristems and embryos. Micropropagation is the vegetative (asexual) propagation of plants under *in vitro* conditions and is widely used for commercial purposes worldwide [1-3].

Plant tissue culture techniques have certain advantages over traditional ones of propagation. These are:

- Thousands of mature plants can be produced in a short time that allows fast propagation of new cultivars,
- Endangered species can be cloned safely,
- Large quantities of genetically identical plants can be produced,
- Plant production is possible in the absence of seeds,
- The production of plants having desirable traits such as good flowers, fruits and odor is possible,
- Whole plants can be regenerated from genetically modified plant cells,
- Disease-, pest- and pathogen-free plants in sterile vessels are produced,
- Plants that their seeds have germination and growing problems such as orchids and nepenthes, can be easily produced,

• Providing infection-free plants for mass production is possible.

Plant tissue culture is based on totipotency which means that a whole plant can be regenerated from a single cell on a growth medium. One of the main objectives of tissue culture studies is to obtain high-frequency shoot regeneration, which is also a prerequisite for an efficient transformation system and a clonal propagation of plants with attractive flowers and fruits in large scale for ornamental purposes. Specially the introduction of foreign genes coding agronomically important traits into plant cells has no meaning unless transgenic plants are regenerated from the genetically modified cell(s).

It is known that some families and genera such as *Solanacea* (*Nicotiana*, *Petunia* and *Datura*), *Cruciferae* (*Brassica* and *Arabidopsis*), *Gesneriaceae* (*Achimenes* and *Streptocarpus*), *Asteraceae* (*Chichorium* and *Chrysanthemum*) and *Liliaceae* (*Lilium* and *Allium*) have a high regeneration capacity while regeneration in some other families such as *Malvaceae* (*Gossypium*) and *Chenopodiaceae* (*Beta*) is difficult. In order to increase the regeneration capacity of explant from the genotype of interest, we have to find answers to such questions as "Why do some genotypes regenerate easily?", "What can be done to increase the regeneration capacity of explant?".

2. Factors affecting explant's regeneration capacity

2.1. Plant material

Plant material is extremely important for the success of tissue culture studies [4]. Factors affecting explant's tissue culture response are (1) genotype, (2) physiological stage of donor plant, (3) explant source, (4) explant age, (5) explant size, (6) explant position in donor plant and (7) explant density. Plant segments used in tissue culture as explant are stem [5], root [6], leaf [7], flower [8], ovule [9], cotyledon and hypocotyl [10, 11]. Such these explants form direct and indirect organs and embriyos. Thin cell layer can also be used as explant in some species [12] while embryos can be successfully used in cereals [13]. Moreover, shoot tips and meristems may give successful results for callus formation and shoot regeneration [14].

2.1.1. Genotype

Regeneration capacity of plants shows a wide range among families, species and even within genotypes from the same species (Figure 1). Generally dicotyledons regenerate more easily than monocots. Plants from some dicotyledon families such as *Solanacea*, *Cruciferae*, *Gesneriaceae*, *Begoniaceae* and *Crassulaceae* have a high regeneration capacity. In general, herbaceous plants regenerate more easily than woody plants such as trees and shrubs [3].

Sugarbeet from *Chenopodiaceae* family is known as a recalcitrant genotype with respect to *in vitro* culture and genetic transformation [15, 16] (Figure 2) while regeneration and transformation are quite easy in tobacco from *Solanaceae* family [17]. It was reported that somatic embryogenesis changed from 0.00% to 77.50% in 14 maize genotypes cultured *in vitro* [18].

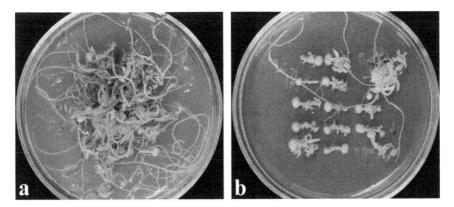

Figure 1. Adventitious shoot regeneration from hypocotyl explants of flax (*Linum usitatissimum* L.) cultivars a. '1886 Sel.' and b. 'Clarck' 4 weeks after culture initiation

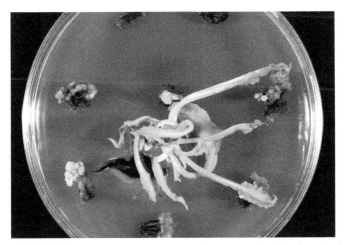

Figure 2. Shoot regeneration from petiole explants of sugar beet (*Beta vulgaris* L.) 4 weeks after culture initiation. Only three explants out of 10 regenerated.

2.1.2. Physiological stage of donor plant

Explants show an ability to express totipotency are the most suitable for tissue culture [19]. Generally, vegetative segments of plants regenerate more easily *in vitro* than generative ones [20]. Explants should be isolated from healthy plants with high cell division for successfull response to tissue culture. On the other hand, regeneration capacity of mature tissues is quite low. In general, young tissues and organs have a high regeneration capacity than the older ones. Regeneration capacity of genotypes increases during flowering period as in *Lunaria annua* [21], *Ranunculus sceleratus* [22]. Although plants in a resting stage (dormant) are generally difficult to culture *in vitro* [3], there are some exceptions as flower stem explants of *Tulipa* form shoots only in dry storage (dormant) stage [23].

2.1.3. Explant source

Plants grown under greenhouse conditions give rise to better results than the ones grown in field conditions [24]. There are huge variations regarding tissue culture response in explants excised from plants grown in field condition depending on wheather conditions during the year [3]. However, the best results are obtained from explants excised from *in vitro* grown seedlings [25].

In a study conducted by Yildiz et al. [25], regeneration capacity of flax (*Linum usitatissimum* L.) hypocotyl explants isolated from *in vitro-* and greenhouse-grown seedlings was compared. Five mm sections of hypocotyl explants from *in vitro*-grown seedlings were directly cultured on MS medium supplemented with 1 mg l^{-1} BAP and 0.02 mg l^{-1} NAA for shoot regeneration while hypocotyls from greenhouse-grown seedlings were surface-sterilized before culture initiation. The highest values with respect to shoot regeneration percentage, shoot number per explant and total shoot number per petri dish were recorded in explants isolated from *in vitro*-grown seedlings of all cultivars.

87.50 shoots were formed over 100.00 in explants isolated from *in vitro*-grown seedlings while only 22.50 shoots were recoved in explants of greenhouse-grown seedlings in cv. '1886 Sel.'. The highest shoot number per explant and total shoot number per petri dish were obtained in explants isolated from *in vitro*-grown seedlings as 14.43 and 126.26, respectively. Specially, total shoot number per petri dish was 42.95 times more in explants of *in vitro*-grown seedlings as 126.26 than in explants excised from greenhouse-grown seedlings as 2.94. It was reported that neither shoot regeneration percentage nor shoot number per explant is lonely an indicator of the success of tissue culture studies but 'total shoot number per petri dish' is a good indicator of the success in both parameters for the genotype of interest [26]. These figures in all the parameters indicated the significance of explant source very well.

2.1.4. Explant age

Regeneration capacity of older plants is often low. As the organ using for explant source gets older, regeneration capacity decreases. An example of differences in regeneration capacity between young and old seedlings that are used as source of explant was flax (*Linum usitatissimum* L.) [27]. In this study, shoot regeneration capacity of hypocotyl explants excised from *in vitro*-grown seedlings at different ages (7, 12 and 17 days) was examined in three flax cultivars. Explants of 7-day-old seedlings gave rise to the highest results with respect to shoot regeneration percentage, shoot number per explant and total shoot number per petri dish. The explants of 7-day-old seedlings were reported to be more vital and well grown (Figure 3).

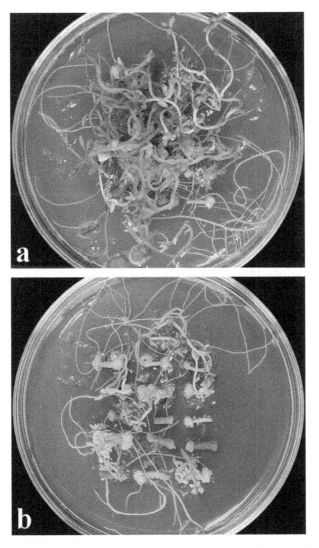

Figure 3. Shoot regeneration from flax (*Linum usitatissimum* L.) cv. '1886 Sel.' hypocotyl explants excised from a. 7-day-old seedling and b. 12-day-old seedling 4 weeks after culture initiation

In the study, it was observed that shoot regeneration percentage, shoot number per explant and total shoot number per petri dish varied excessively among explants from different seedling ages. All these three parameters, which were the highest in explants of 7-day-old seedling were reduced significantly in explants of 12- and 17-day-old seedlings. Results clearly showed that there were statistically significant differences in all parameters examined in all cultivars among explants at different ages.

2.1.5. Explant size

It is so difficult to obtain a successful tissue culture response from small parts such as cells and meristems than from larger parts such as leaves and hypocotyls due to their limited nutrients and hormone reserves. Larger explants having a big amount of nutrition reserves such as tubers and bulbs can easily regenerate *in vitro* and are less dependent on nutrients and hormones in growth medium [3].

2.1.6. Explant position in donor plant

In vitro growth of explants can be affected depending on the place from where they are excised. For instance, the higher parts of donor plant are older than the lower parts. Evers [28] has reported that shoot initials from lower parts of *Pseudotsuga menziesii* developed better under *in vitro* conditions. Explants excised from the base produce adventitious bulbs more easily than the top parts of the bulbs [3].

In a study conducted with flax, tissue culture response of hypocotyl explants at different positions was evaluated. Hypocotyls were classified from where they were excised as top, medium and low. Top part was just below the cotyledon leaves while lower one was close to bottom (Figure 4). Fifteen hypocotyls from 7-day-old *in vitro*-grown seedlings were cultured on MS medium 1 mg l^{-1} BAP and 0.02 mg l^{-1} NAA for 4 weeks.

Explants excised from top (just below cotyledon leaves) part of the seedling gave rise to the highest results with respect to shoot regeneration percentage, shoot number per explant and total shoot number per petri dish. Results were getting lower in the explants excised from medium part of the seedling. And the lowest values were recorded in the explants from lower part (Unpublished study results) (Figure 5).

The highest values in shoot regeneration percentage, shoot number per explant and total shoot number per petri dish were recorded in the explants excised from the top of the seedling as 97.78%, 5.17 and 76.00, respectively. These figures were the lowest in hypocotyl segments in the lower part of the seedling as 66.67%, 2.62 and 26.33, respectively. There were huge difference between the results of explants excised from the top and the low parts of the seedling. Shoot regeneration percentage was 97.78% in explants of the top part of the seedling while it was 66.67%. That is 97.78 explants regenerated over 100.00 in the top part of the seedling.

Mean shoot number per explant was recorded as 5.17 in explants excised from the top parts of the seedling while it was 2.62 in explants isolated from the low part of the seedling. Similar results were obtained in total shoot number per petri dish. Shoot number per explant and total shoot number per petri dish are the main indicators of success in plant tissue culture. Especially, after transformation studies via *Agrobacterium tumefaciens*, regeneration capacity of tissue decreases siginificantly due to plant defense mechanism against pathogen attact. That is why, the number of shoots regenerated should be as much as possible.

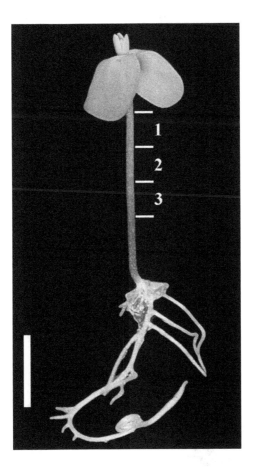

Figure 4. Seven-day-old flax (*Linum usitatissimum* L.) cv. 'Madaras' seedling and the position from which explants were excised 1. top, 2. medium and 3. low. Bar = 1.0 cm

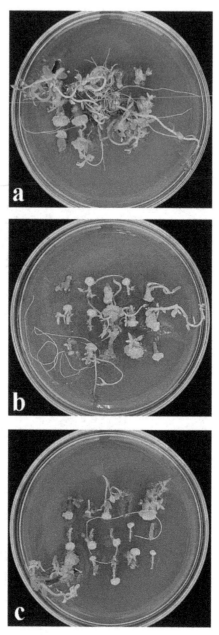

Figure 5. Tissue culture response of hypocotyl explants excised from different positions of flax (*Linum usitatissimum* L.) cv. 'Madaras' seedling a. top, b. medium and c. low

2.1.7. Explant Density

There are many studies about competition among the plants in field conditions. Plant leaves compete for irradiance, and roots for water and nutrients [29, 30]. High plant density was accepted as a biotic stress factor [31, 32]. Stoffella and Bryan [33] have reported that plant density has an effect on plant development and yield of many vegetable crops. A linear increase has indicated in fruit yield when plant density is increased [34-36]. Abubaker [37] has noted that the highest planting density gave rise to the lowest yield in bean due to the high competition among plants for water and minerals. Asghari et al. [38] have reported that the chicory plant increased root diameter for increased absoption of water under high density by high competition between plants.

Since a plant is a unity of cells and tissues, behavior of cells and tissues as the smallest unit of the organism, represents the plant's response against any factors arising from nearby environment. Competition and stress are the brother concepts that should be taken into account to increase the success of tissue culture studies. Explants under *in vitro* conditions compete with each other for a constant amount of water and nutrients in growth medium. It was revealed that competition was one of the most important factors increasing explant's regeneration capacity [32].

Yildiz [32] has reported that encouraging hypocotyl explants of flax (*Linum usitatissimum* L.) for competition by decreasing the culture distances among them increased shoot regeneration capacity remarkably till a certain point from where stress initiated and significant decreases in all parameters were observed (Figure 6 and Figure 7). For shoot regeneration, hypocotyls were cultured in a petri dish (90 × 90 mm) at 0.5 x 0.5, 1.0 x 1.0, 1.5 x 1.5 and 2.0 x 2.0 cm distances.

Results clearly showed that there were statistically significant differences in fresh and dry weights in all cultivars among explants cultured at different culture distances. The highest fresh and dry weights per explant of all cultivars were obtained from 2.0 x 2.0 cm distance of hypocotyls and they decreased by decreasing distances. These findings were supported by Gersani et al. [39] and Maina et al. [40] who reported that plants grown alone produce more biomass or yield than those grown with the others. It could be concluded that the decreased distance in which explants were cultured induced stress caused likely by the deficiency of water, sucrose and nutrients. Increases in the fresh and dry weights were chiefly due to an increase in the absorption of water and other components from the basal medium [41]. It was stated that the fresh weight increase was mainly due to cell enlargement by water absorption [42] and increase in dry weight was closely related to cell division and new material synthesis [43].

The highest shoot number per hypocotyl (20.70 in 'Madaras', 14.57 in '1886 Sel.' and 17.40 in 'Clarck') and the highest shoot length (3.10, 2.14 and 2.09 cm, respectively) were obtained at 1.0 x 1.0 cm distance in all cultivars studied. It is thought that competition among explants cultured at 1.0 x 1.0 cm distance encouraged them to give higher results than at the other distances.

Plant growth regulators are perhaps the most important components affecting shoot regeneration capacity of explants [44]. In tissue culture studies, correct combinations of auxins and cytokinins have been tried to be determined for high frequency shoot regeneration for the explant [26]. This study has shown that determination of optimum levels of auxins and cytokinins in growth medium is not the only way of increasing shoot regeneration capacity, but also shoot regeneration frequency of explants could be increased simply by encouraging explants into competition.

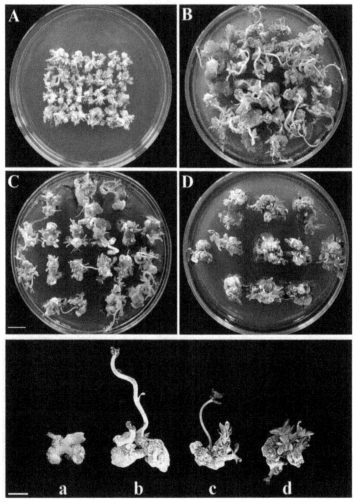

Figure 6. Development of explants cultured at different distances 0.5 x 0.5 cm (A, a), 1.0 x 1.0 cm (B, b), 1.5 x 1.5 cm (C, c) and 2.0 x 2.0 cm (D, d); six weeks after culture initiation. Bar is 1.0 cm for petri dishes and 0.5 cm for shoot regeneration

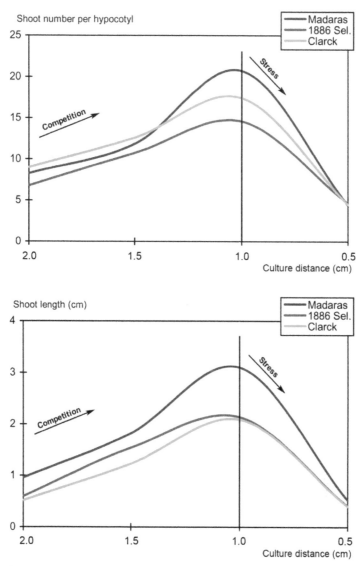

Figure 7. Competition-stress curve of flax cultivars 'Madaras', '1886 Sel.' and 'Clarck' with respect to shoot number per hypocotyl and shoot length

2.2. Surface-sterilization process

All tissue culture studies which aim to obtain high-frequency shoot regeneration, which is also a prerequisite for an efficient transformation system, should be performed under sterile conditions. Explant health is the main factor determining regeneration capacity. Viability of explant and the seedling from which the explant is excised, are very important for high-frequency shoot regeneration [45]. The most important treatment prior to culture initiation is perhaps surface-sterilization of the explant. Since *in vitro* conditions provide bacteria and fungi with an optimal growth environment, unsuccessful sterilization hinders the progress of tissue culture studies. Surface-sterilization process aims to eliminate all microorganisms that can easily grow under *in vitro* conditions; on the other hand, it should guarantee the explant's viability and regeneration capacity, which are known to be affected by the concentration, application period [46] and temperature [45] of disinfectant. Since direct contact of explant with disinfectant during the sterilization process may have a severe effect on regeneration capacity of the tissue [45], using aseptic tissues as source of explant is highly recommended [47, 48].

A wide range of surface disinfectants, such as ethanol, hydrogen peroxide, bromine water, mercuric chloride, silver nitrate, and antibiotics are used for surface-sterilization; however sodium hypochlorite (NaOCl) has been most widely used. NaOCl is highly effective against all kinds of bacteria, fungi, and viruses [49-52]. Moreover, NaOCl has a strong oxidizing property which makes it highly reactive with amino acids [53, 54], nucleic acids [55], amines, and amides [56, 57]. The general reaction between amino acids and NaOCl produces the respective aldehyde, NH_4Cl and CO_2 [54].

2.2.1. Effect of sodium hypochlorite solutions at different concentrations and application periods

In vitro seed germination, seedling growth and the viability of the tissue were negatively affected by sodium hypochlorite (NaOCl) at high concentrations [45, 58, 59] while it is uneffective for sterilization of tissues at low concentrations. The negative effects of NaOCl concentration become more severe with increasing application period. Since regeneration capacity of the tissue is negatively affected by higher concentrations and longer application periods of disinfectants [3, 46], sterilization process under *in vitro* conditions should aim to use the lowest concentration of disinfectant for the shortest time.

In the study aiming to evaluate the effects of NaOCl solutions used for sterilization on *in vitro* seed germination and seedling growth in *Lathyrus chrysanthus* Boiss., the best results were obtained from 3.75% NaOCl concentration and 15 min. application period for all parameters examined [60]. Seedborne contamination increased gradually by decreasing concentrations and application periods of NaOCl below 3.75% and 15 min. Dramatic decreases were observed at 5.00% NaOCl concentration in all cases. At this concentration, NaOCl showed deleterious effect on the embryo of the seed. Seed germination decreased to 65.18% when NaOCl concentration increased to 5.00% from 3.75% at 15 min. application period. Seedling growth from seeds sterilized with 3.75% NaOCl concentration for 15 min. were observed growing faster than that of sterilized with other concentrations and application periods of NaOCl (Figure 8). By increasing NaOCl concentration to 5.00%, seedling length decrased to 3.45 cm from 3.90 cm. Higher results in seedlings grown from seeds sterilized with 3.75% NaOCl concentration for 15 min. could be caused by higher tissue water content as reported that *in vitro* explant growth and plantlet establishment have been affected significantly by tissue water content [41].

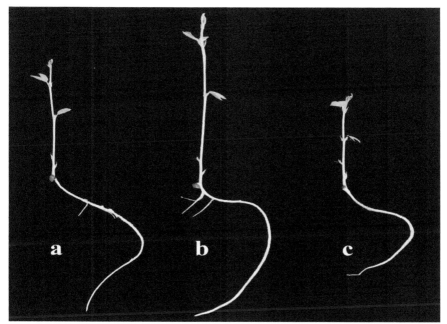

Figure 8. *In vitro* seedling growth from *Lathyrus chrysanthus* seeds sterilized with (a) 2.50%,   (b) 3.75% and (c) 5.00% NaOCl for 15 min.

2.2.2. Effect of sodium hypochlorite solutions at different temperatures

It was firstly reported that besides concentration and application period, temperature of NaOCl was also one of the most important factors affecting *in vitro* seed germination, seedling growth and explant's regeneration capacity [45].

At the 2.00% NaOCl concentration using for surface-sterilization of flax (*Linum usitatissimum* L.) seeds, when the temperature of NaOCl was set below 10°C, bacterial and fungal contamination was observed. However, increases in NaOCl temperature above 10°C resulted in dramatic decreases in seed germination, seedling growth, hypocotyl and root lengths. When flax seeds were surface-sterilized with 3.00% and 4.00% NaOCl concentrations at 30°C, seed germination, seedling growth, hypocotyl and root lengths decreased dramatically. Decreases in all parameters in NaOCl temperature above 10°C could be the fact that disinfection activity of NaOCl increases [61] and disinfectant penetrates more easily through the seed coat [62]. Higher NaOCl temperatures resulted in morphologically abnormal seedlings with stunted hypocotyls and roots (Figure 9).

Yildiz and Er [45] reported that increasing disinfectant temperature using for surface-sterilization of flax seeds to obtain sterile *in vitro* seedlings from which hypocotyls were isolated, reduced shoot regeneration significantly.

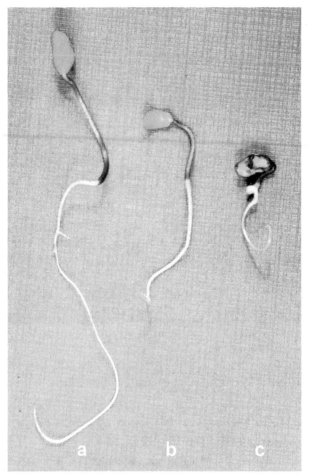

Figure 9. *In vitro* seedling growth in flax from seeds sterilized with 2.00% NaOCl at temperatures of (a) 10°C, (b) 20°C and (c) 30°C

Regeneration capacity of the explant is negatively affected by higher concentrations, application periods and temperatures of disinfectant used for surface-sterilization. In the sterilization process of the tissues, the concentration, application period and temperature of NaOCl solutions are closely related to each other and they should be considered together. Direct contact of the tissue with disinfectant during the sterilization process may have a severe effect on the viability and regeneration capacity depending on concentration, temperature and application periods of disinfectant [45, 46]. In addition to this common knowledge, not only seed germination and seedling growth are directly affected by sterilization process, but also regeneration capacity of explants and health of regenerated shoots are indirectly influenced significantly in plant tissue culture studies. That means NaOCl affects the success of *in vitro* studies, from *in vitro* seed germination and seedling establishment to regeneration capacity of the tissue and recovery of plantlets.

2.3. Culture medium

The composition of growth medium is an important factor affecting growth and morphogenesis of plant tissues. Plant tissue culture medium consists of macronutrients, micronutrients, vitamins, amino acids or other nitrogen supplements, carbon sources, organic supplements, solidifying agents and growth regulators. Murashige and Skoog [63] are the most commonly used medium in plant tissue culture. The B5 [64], N6 [65] and Nitsch and Nitsch [66] (NN) have been widely used for many plant species. Moreover, for culture of woody species, the Driver/Kuniyuki walnut medium (DKW) [67] and the WPM medium [68] are used. The growth medium is selected for the purpose of tissue culture and for the plant species [69].

Yildiz et al. [27] have conducted a study to evaluate the effects of two different growth medium (MS and B5) and two gelling agents (Agar and Phytagel) on the regeneration capacity of flax (*Linum usitatissimum* L.) hypocotyl explants of three cultivars namely 'Madaras', '1886 Sel.' and 'Clarck'. Results showed that MS (Murashige and Skoog) growth medium and Agar as gelling agent gave rise to the highest results with respect to shoot regeneration percentage, shoot number per explant and total shoot number per petri dish in all cultivars studied (Table 1).

Cultivars	Shoot regeneration (%)				Shoot number per explant				Total shoot number per petri dish			
	1	2	3	4	1	2	3	4	1	2	3	4
'Madaras'	97.22	91.67	80.56	91.67	7.41a	4.70b	2.56cd	2.50cd	86.67a	56.34bc	30.67cde	30.00cde
'1886 Sel.'	100	100	100	100	6.72a	5.56ab	4.14cd	3.95cd	80.34a	66.67ab	49.67cd	47.34cd
'Clarck'	97.22	88.89	88.89	94.44	4.81a	3.02b	2.59bc	2.34bc	57.67a	35.00b	31.00bc	28.00bcd

1. MS medium and Agar; 2. MS medium and Phytagel; 3. B5 medium and Agar; 4. B5 medium and Phytagel
Values in a row for each cultivar followed by the different letters are significantly different at the 0.01 level

Table 1. The effect of different growth medium and gelling agents on shoot regeneration capacity of flax hypocotyl explants

2.4. Culture conditions

After explants placed on growth medium for different purposes, they should be cultured in culture rooms where the environmental factors such as temperature and light are controlled. Different species may need different environmental conditions for successful culture.

Lighting in culture rooms is realized by fluorescent tubes. Control equipments of tubes should be set up outside the culture room. Otherwise they may cause over heating inside the room and in that case extra cooling is necessary. Due to light sources inside the culture room, there should be an efficient cooling system to maintain constant temperature conditions. Fluorescent tubes can be installed under the shelves, above the cultures which provide a more uniform irradiation for the cultures. Although 16h light and 8h dark photoperiod is usually used, there may be some differences according to species grown under long-day or short-day conditions.

The temperature in culture room is so important for successful tissue culture. Temperature variation in culture room should be as small as possible and generally ±1ºC is allowed. Otherwise, changing temperature regime causes stress in cultures which is one of the main reasons of unsuccess. That is why, working with many culture rooms are recommended instead of working only with one.

3. Treatments increasing explant's regeneration capacity

3.1. Increasing tissue water content

Water is the source of life on earth. Life in a large proportion of terrestrial ecosystems is limited by water availability. The water content of an actively growing plant can be as much as 95% of its live weight. A plant requires water as an essential ingredient of photolysis, the photochemical stage of photosynthesis where water is split using light energy. Neither carbon dioxide nor oxygen required for photosynthesis is usable by plant unless it is in solution in water. Therefore, water is the key to plant's survival and growth. Water is also an excellent solvent. The substances (solutes) that become dissolved in water in plants include mineral ions such as potassium (K^+), sugars (glucose and sucrose), and amino acids, main components of proteins.

The reduction in growth, yield and quality by water stress has been well recognized in field conditions [70-71]. Germination and seedling establishment guarantee plant survival and are very important phases of plant life. Germination rate decreases with decreasing external water potential and for each species there is a critical value of water potential below which germination will not occur [72].

Yildiz and Ozgen [41] have reported that tissue water content affected explant's shoot regeneration capacity significantly. In the study, water-treated and non-water treated hypocotyl explants of three flax cultivars were compared with regard to fresh and dry weights, shoot regeneration percentage, shoot number per explant, shoot length and total shoot number per petri dish. Some hypocotyls were submerged in sterile distilled water with a gentle shaking for 20 min before placing on regeneration medium, while others were directly cultured on MS medium supplemented with 1 mg l^{-1} 6-benzylaminopurine (BAP) and 0.02 mg l^{-1} naphthaleneacetic acid (NAA) for regeneration. Results clearly showed that there were sharp and statistically significant differences in all cultivars between water-treated and non-water-treated explants concerning all characters examined (Table 2, Figure 10).

Possibly, pretreatment of explants with water softened the epidermis layer and increased the permeability which caused to high tissue metabolic activity by increasing water and hormone uptake from the medium. Thus, increase in the fresh and dry weights of water-treated hypocotyl explants at the end of culture were chiefly due to an increase in the absorption of water and other components from the basal medium via high permeable epidermis membrane. In the study, non-water-treated explants were found smaller than water-treated ones in all cultivars. Dale [42] stated that the fresh weight increase is mainly due to cell enlargement by water absorption, cell vacuolation, and turgor-deriven wall expansion. The increase in dry weight was closely related to cell division and new material synthesis [43]. Dry weight increase of water-treated explants is due to an increase in carbohydrate metabolism resulting from increased water uptake. On the other hand, lower levels of all parameters of non-water-treated explants were directly due to a decreased water uptake from the environment and consequently, a reduced mobilization of plant growth regulators. Hsiao [73] has reported that the inhibition of growth under water stress conditions is the result of inhibition of cell division, cell elongation or both. Osmotic water absorption affects cell elongation. It has been suggested that osmotic stress modifies the biochemical changes taking place in the cell wall during growth thereby preventing extension [74]. The primary action of osmotic inhibition is retardation of water uptake which is vital for germination and growth [75]. It has been stated that water stress alters the level of plant hormones [76].

Cultivar	Hypocotyl				Shoot regeneration (%)	
	Fresh weight (g)		Dry weight (g)			
	WT	NWT	WT	NWT	WT	NWT
'Madaras'	0.430±0.047	0.216±0.019	0.034±0.002	0.025±0.002	100±0.000	75.00±2.937
'1886 Sel.'	0.343±0.011	0.231±0.013	0.029±0.013	0.021±0.001	100±0.000	73.34±4.051
'Clarck'	0.396±0.013	0.192±0.025	0.031±0.001	0.019±0.002	100±0.000	78.33±2.305
Cultivar	Shoot number per hypocotyl		Shoot length (cm)		Total shoot number per petri dish	
	WT	NWT	WT	NWT	WT	NWT
'Madaras'	12.17±0.210	8.00±0.328	0.56±0.043	0.25±0.018	182.50±3.142	119.88±4.90
'1886 Sel.'	11.20±0.114	8.15±0.367	0.58±0.029	0.22±0.009	167.88±1.712	122.25±5.487
'Clarck'	10.83±0.265	5.26±0.491	0.60±0.043	0.27±0.020	162.50±3.974	78.88±7.392

In all cases, the values for WT and NWT were significantly different at 0.01 level with the exception of dry weight of hypocotyls of Madaras which is different at 0.05

Table 2. Adventitious shoot regeneration from water-treated (WT) and non-water-treated (NWT) hypocotyls of three flax cultivars 6 weeks after culture initiation on MS medium containing 1 mg l⁻¹ BAP and 0.02 mg l⁻¹ NAA.

Treatment of explants with water before culture initiation increased permeability of the epidermis layer and the tissue's water content and so enabled water, all solutes and plant growth regulators to transfer into the tissue more easily, providing all cells with a high regeneration capacity and consequently increasing explant's tissue culture response.

Shoots regenerated from water-treated and non-water-treated explants were rooted on MS medium containing 3 mg l⁻¹ IBA for 3 weeks. The best results were obtained in the shoots regenerated from water-treated explants (Figure 10c). Sharp and dramatic differences, which were all statistically significant at the 0.01 level, were observed in all parameters between the shoots regenerated from water-treated and non water-treated explants (Table 3). Similar effects of water treatment were also noted in rooting stage. This means that shoots regenerated from water-treated explants were more capable of establishing new plantlets than the ones grown from non-water-treated explants.

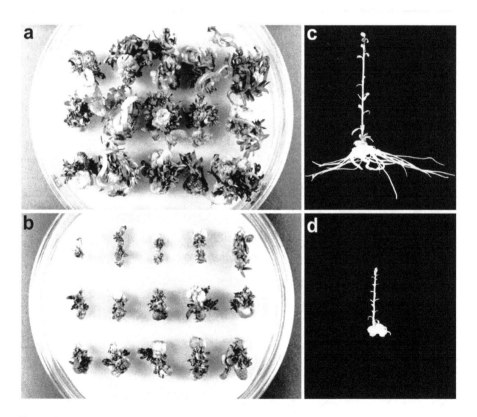

Figure 10. *In vitro* shoot regeneration from water-treated (**a**) and non-water-treated (**b**) hypocotyls of *Linum usitatissimum* cv. '1886 Sel.' from 6-week-old culture. *In vitro* rooting and development of shoots regenerated from water-treated (**c**) and non-water-treated (**d**) hypocotyls of *Linum usitatissimum* cv. '1886 Sel.' 3 weeks later.

Cultivar	Shoot length (cm)		Number of roots		Mean length of each root (cm)	
	WT	NWT	WT	NWT	WT	NWT
'Madaras'	3.02±0.225	1.65±0.209	10.20±1.519	6.50±1.053	1.58±0.156	1.16±0.192
'1886 Sel.'	3.98±0.220	1.28±0.185	21.31±2.121	7.19±1.342	1.92±0.144	0.82±0.076
'Clarck'	4.81±0.396	2.10±0.156	29.00±2.887	14.63±1.812	2.33±0.223	1.56±0.143

In all cases, the values for WT and NWT were significantly different at 0.01 level with the exception of mean length of each root of shoots of Madaras which is different at 0.05

Table 3. *In vitro* root development of shoots regenerated from water-treated (WT) and non-water-treated (NWT) hypocotyl explants on rooting medium enriched with 3 mg l^{-1} IBA 3 weeks after culture initiation

3.2. Regulating osmotic pressure of explant

In another study conducted by Yildiz et al. [77], pretreated and non-pretreated hypocotyl explants of three flax cultivars ('Omega', 'Fakel' and 'Ariane') were cultured for adventitious shoot regeneration. Two different pretreatment applications were compared to the conventional regeneration protocol with respect to hypocotyl fresh and dry weights, shoot regeneration percentage, shoot number per hypocotyl, shoot length and total chlorophyll content. In the 1st and 2nd pretreatment applications, hypocotyl explants were kept in sterile cabin under air flow for 30 min. in order to make them dry as reported by Christmann et al. [78] to decrease the tissue water content and to gain explants the ability of uptaking increased amount of water, all solutes and plant growth regulators by using osmotic pressure in consequent applications. Then explants were treated with MS solution containing 1 mg l^{-1} BAP and 0.02 mg l^{-1} NAA for 15 min. Finally all explants were cultured on MS medium without growth regulators in the 1st pretreatment application and on MS medium enriched with 1 mg l^{-1} BAP and 0.02 mg l^{-1} NAA in the 2nd pretreatment application. It was expected that by immersing explants into liquid regeneration medium after drying enabled all cells to absorb more growth regulators along with water in both pretreatment applications. However, only in 2nd pretreatment application, explants were cultured on MS medium containing 1 mg l^{-1} BAP and 0.02 mg l^{-1} NAA which means that tissues maintained uptaking increased water and growth regulators from the regeneration medium which led to higher results in all parameters studied as reported by Yildiz and Ozgen [41]. Likewise, Okubo et al. [79] have reported that endogenous hormone levels of tissue affected regeneration capacity *in vitro* significantly. Fatima et al. [80] have also noted that internal factors such as chemicals and mineral nutrients affect *in vitro* plant growth. The required amount of exogenous plant growth regulators for cultured tissues depends on the endogenous levels plant tissues have [80]. It was first reported that keeping explants in sterile distilled water for 20 min. before culture on MS medium enriched with 1 mg l^{-1} BAP and 0.02 mg l^{-1} NAA increased the regeneration capacity of hypocotyls of flax tremendously by increasing permeability of the epidermis layer and tissue's water content and enabling water, all solutes and growth regulators to transfer into the tissue more easily [41].

According to the results, there were statistically significant differences among pretreated and non-pretreated hypocotyls in all cultivars (Table 4).

Cultivar	Pre. App.	Hypocotyl		Shoot regeneration (%)*	Shoot number per hypocotyl	Shoot length (cm)	Total chlorophyll content (µg/g fresh tissue)
		Fresh weight (g)	Dry weight (g)				
'Omega'	1	0.25"±0.020c	0.014 ±0.0019b	82.40±1.07b	6.76±0.46b	1.93±0.13b	217.1±10.40c
	2	0.48±0.023a	0.034 ±0.0023a	100.00±0.00a	11.38 ±0.69a	2.82±0.14a	380.6±26.91a
	3	0.37±0.017b	0.018 ±0.0017b	90.00±5.77ab	7.99±0.74b	1.51±0.17b	286.2±10.45b
'Fakel'	1	0.21±0.027b	0.016 ±0.0030b	72.00 ±7.53b	6.42±0.19c	1.05±0.11b	197.0±15.40c
	2	0.42±0.006a	0.032 ±0.0046a	100.00±0.00a	8.89±0.37a	1.61±0.18a	316.5±14.37a
	3	0.31±0.045b	0.024 ±0.0031ab	80.60±7.45a	7.49±0.10b	0.96±0.10b	252.1±9.89b
'Ariane'	1	0.19±0.024b	0.014 ±0.0023b	46.23 ±6.20c	4.26±0.18b	1.27±0.06b	192.0±11.25c
	2	0.36±0.055a	0.030 ±0.0012a	100.00 ±0.00a	6.64±0.25a	2.00±0.13a	346.0±18.62a
	3	0.26±0.026ab	0.019 ±0.0018b	65.35 ±3.76b	4.59±0.07b	1.06±0.09b	268.2±24.10b

Pretreatment applications;

1. Hypocotyls were waited for 30 min. in sterile cabin under air flow and for 15 min. in solution containing 1 mg l^{-1} BAP+0.02 mg l^{-1} NAA and finally cultured on MS0 medium

2. Hypocotyls were waited for 30 min. in sterile cabin under air flow and for 15 min. in solution containing 1 mg l^{-1} BAP+0.02 mg l^{-1} NAA and finally cultured on MS medium containing 1 mg l^{-1} BAP+0.02 mg l^{-1} NAA

3. Hypocotyls were cultured directly on MS medium containing 1 mg l^{-1} BAP+0.02 mg l^{-1} NAA (non-pretreated)

*The values represent mean ± standard error of the mean

**Values within a column for each cultivar followed by different letters are significantly different at the 0.01 level.

Table 4. Tissue culture response from pretreated and non-pretreated hypocotyls of three flax cultivars 6 weeks after culture initiation

The highest results in both fresh and dry weights of hypocotyls of all cultivars were obtained from 2nd pretreatment application. Scores of fresh and dry weights were followed by non-pretreated hypocotyls. The lowest results were recorded from 1st pretreatment application in all cultivars studied (Table 4). From the results, it could be concluded that increases in the fresh and dry weights were chiefly due to an increase in the absorption of water and growth regulators from the medium where explants were first pretreated and then cultured. When the results of 2nd pretreatment application were examined, it could be easily seen that culturing explants on MS medium containing 1 mg l^{-1} BAP and 0.02 mg l^{-1} NAA after treating them with liquid MS medium supplemented with 1 mg l^{-1} BAP and 0.02 mg l^{-1} NAA clearly enriched the tissue's growth regulators level which caused to higher fresh and dry weights.

The results related to shoot regeneration percentage indicated that the lowest results were obtained from the 1st pretreatment application in all cultivars. Hypocotyl explants formed roots and fewer callus in the 1st pretreatment application than the others. All explants regenerated successfully in the 2nd pretreatment application and consequently shoot regeneration percentage was recorded as 100% in all cultivars studied (Table 4, Figure 11).

The highest results in shoot number per hypocotyl and shoot length were obtained from 2nd pretreatment application in all cultivars studied. The highest shoot number per hypocotyl was recorded as 11.38 in 'Omega', 8.89 in 'Fakel' and 6.64 in 'Ariane'. The highest scores related to

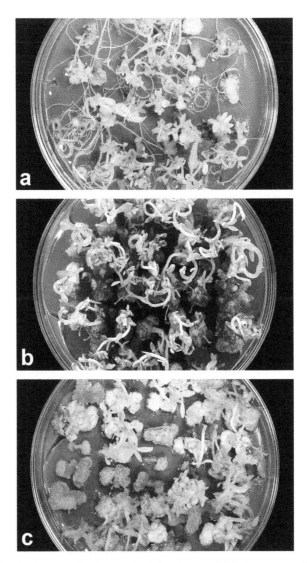

Figure 11. *In vitro* shoot regeneration from pretreated and non-pretreated hypocotyls of flax (*Linum usitatissimum*) cv. 'Omega'. **(a)** 1st pretreatment application: Hypocotyls were waited for 30 min. in sterile cabin under air flow and treated with solution containing 1 mg/l BAP and 0.02 mg/l NAA for 15 min. and finally cultured on MS0 medium, **(b)** 2nd pretreatment application: Hypocotyls were waited for 30 min. in sterile cabin under air flow and treated with solution containing 1 mg/l BAP and 0.02 mg/l NAA for 15 min. and finally cultured on MS medium containing 1 mg/l BAP+0.02 mg/l NAA, **(c)** Non-pretreatment application: Hypocotyls were directly cultured on MS medium containing 1 mg/l BAP+0.02 mg/l NAA

shoot length were 2.82, 1.61 and 2.00 cm in 'Omega', 'Fakel' and 'Ariane', respectively. Shoot regeneration capacity of hypocotyls increased significantly in 2nd pretreatment application. The best results in total chlorophyll content were obtained from 2nd pretreatment application in all cultivars. The highest scores of total chlorophyll content were recorded as 380.6 µg/g fresh tissue in 'Omega', 316.5 µg/g fresh tissue in 'Fakel' and 346.0 in 'Ariane' (Table 4). The explants to which 2nd pretreatment application was carried out were more vital and well-grown and more capable of regeneration (Figure 11). Emerson [81] reported that there is a close relationship between photosynthesis and chlorophyll content. Chlorophyll content of leaf is considered as a sign of photosynthetic capacity of tissues [81-84] which plays a critical role in plant growth and development [85] and its amount changes under stress conditions [86-88]. Gireesh [89] has reported that chlorophyll can be used to measure growth.

From the results, it could be concluded that the lower levels of all parameters recorded in the 1st and 3rd pretreatment applications were directly due to a decreased uptake of water and growth regulators from the medium. Tissue culture response has been affected significantly by tissue water content [41]. Treatment of explants with liquid MS medium containing 1 mg l⁻¹ BAP and 0.02 mg l⁻¹ NAA for a while before culture initiation enabled water, all solutes and plant growth regulators to transfer into the tissue much more, providing all cells with a high regeneration capacity and consequently increasing explant's tissue culture response.

4. Conclusion

Plant tissue culture techniques help us to propagate plants vegetatively in a large amount starting from small parts of a tissue and by using the potential of known as totipotency, to form a whole, fertile plant. Plant tissue culture studies are performed on an artificial growth medium under sterile conditions. Explants regenerate shoots and roots, and consequently whole fertile plants under certain cultural conditions. Tissue culture studies aim to obtain high-frequency shoot regeneration, which is also a prerequisite for an efficient transformation system and a clonal propagation of plants. The introduction of foreign genes coding agronomically important traits into plant cells has no meaning unless transgenic plants are regenerated from the genetically modified cell(s). Therefore, using tissues having high regeneration capacity is extremely important. Regeneration capacity of cells or tissues to be used in transformation studies, affects the success of genetic transformation significantly. The types and concentrations of plant growth regulators in plant cell culture significantly affect growth and morphogenesis. In order to obtain high frequency adventitious shoot regeneration for related genotype, correct concentrations and combinations of auxins and cytokinins should be determined. However, determining the explant type, and correct concentrations and combinations of growth regulators is not enough for high frequency shoot regeneration. Since every cell has an ability of forming a whole fertile plant under *in vitro* conditions, shoot regeneration frequency can always be higher than we obtain in theory. Many factors affecting regeneration capacity of explant are not found out yet. For instance, a recently reported technique utilizing competition among explants is very effective to increase shoot regeneration capacity. Thus, unknown factors affecting regeneration capacity of explants should be determined in order to increase the success of tissue culture studies.

Author details

Mustafa Yildiz*
Department of Field Crops, Faculty of Agriculture, University of Ankara, Diskapi, Ankara, Turkey

5. References

[1] Huetteman CA, Preece JE. Thidiazuron: a potent cytokinin for woody plant tissue culture. Plant Cell Tissue Organ Culture 1993; 33: 105-119.

[2] Mantell SH, Haque SQ, Whitehall AP. Clonal multiplication of Dioscorea alata L. and Dioscoren rotiindata Poir. yams by tissue culture. Journal of Horticultural Science 1978; 53(2): 95-98.

[3] Pierik RLM. *In vitro* Culture of Higher Plants. Martinus Nijhoff Publishers, Dordrect; 1987.

[4] Tisserat B. Embryogenesis, Organogenesis and Plant Regeneration. In: Dixon RA. (ed). Plant Cell culture: A Practical Approach. IRL Press, Oxford, Washington DC; 1985, p79-105.

[5] Skoog F, Tsui C. Chemical control of growth and bud formation in tobacco stem segments and callus cultured *in vitro*. American Journal of Botany 1948; 35: 782-787.

[6] Earle ED, Torrey JG (1965) Morphogenesis in cell colonies grown from *Convolvulus* cell suspensions plated on synthetic media. American Journal of Botany, 52: 891-899.

[7] Gürel E. Transferring an antimicrobial gene into *Agrobacterium* and tobacco. May 20-22, 1998, 2nd International Kizilirmak Fen Bilimleri Congress, Kırıkkale University, Kirikkale, Turkey.

[8] Kaul K, Sabharwal PS. Morphogenetic studies on Haworthia: Establishment of tissue culture and control of differentiation. American Journal of Botany 1972; 59: 377-385.

[9] Gürel S, Gürel E, Kaya Z. Ovule culture in sugar beet (*Beta vulgaris* L.) breeding lines. May 20-22, 1998, 2nd International Kızılırmak Fen Bilimleri Congress, Kırıkkale University, Kirikkale, Turkey.

[10] Gürel E, Kazan K. Development of an efficient plant regeneration system in sunflower (*Helianthus annuus* L.) Turkish Journal of Botany 1998; 22: 381-387.

[11] Yildiz M. 2000. Adventitious shoot regeneration and Agrobacterium tumefaciens-mediated gene transfer in flax (Linum usitatissimum L.). PhD thesis. University of Ankara, Graduate School of Natural and Applied Sciences, Department of Field Crops. Ankara; 2000.

[12] Tran Thanh Van K. Control of morphogenesis in *in vitro* cultures. Annual Review of Plant Physiology 1998; 32: 291-311.

[13] Sommer HE, Brown CK, Kormanik PP. Differentiation of plantlets in longleaf pine (*Pinus palustris* Mill.) tissue cultured *in vitro*. Botanical Gazette 1975; 136: 196-200.

[14] Vasil IK, Vasil V. Regeneration in Cereal and Other Grass Species. In: Vasil IK (ed.) Cell Culture and Somatic Cell Genetics of Plants. Academic Pres, New York; 1986. P121-150.

* Corresponding Author

[15] Tetu T, Sangwan RS, Sangwan-Norreel BS. Hormonal control of organogenesis and somatic embryogenesis in *Beta vulgaris* callus. Journal of Experimental Botany 1987; 38: 506-517.

[16] Krens FA, Jamar D. The role of explant source and culture conditions on callus induction and shoot regeneration in sugarbeet (*Beta vulgaris* L.). Journal of Plant Physiology 1989; 134: 651-655.

[17] Budzianowska A. In vitro cultures of tobacco and their impact on development of plant biotechnology. Przeglad Lekarski 2009; 66(10): 890-893.

[18] Emeklier Y, Ozcan S, Avcı Birsin M, Mirici S, Uranbey S. Studies on in vitro somatic embryogenesis in maize. Ankara University, Research Fund Project, Ankara, Turkey; 1999.

[19] Mantell SH, Matthews JA, McKee RA. Principles of plant biotechnology. An introduction to genetic engineering in plants. Blackwell Scientific Publications, Oxford; 1985.

[20] Robb Sheila M. The Culture of Excised Tissue 3*Lilium speciosum* Thun. Journal of Experimental Botany, 1957, 8(3): 348-352.

[21] Pierik RLM. Meded. Landbouwhogeschool Wageningen 1967; 67(6): 1-71.

[22] Konar RN, Nataraja K. Morphogenesis of isolated floral buds of *Ranunculus sceleratus* L. in vitro. Acta Botanica Neerlandica 1969; 18: 680-699.

[23] Wright NA, Alderson PG. The growth of tulip tissues in vitro. Acta Horticulturae 1980; 109:263–270.

[24] Gürel E, Türker AU (2001) Organogenesis. In: Babaoğlu M, Gürel E, Özcan S (eds.) Plant Biotechnology, Tissue Culture and Applications. Selcuk University Publications, Konya, Turkey; 2001. P36-70.

[25] Yıldız M, Ozcan S, Er C. The effect of different explant sources on adventitious shoot regeneration in flax (*Linum usitatissimum* L.). Turkish Journal of Biology 2002; 26: 37-40.

[26] Yıldız M, Saglik Ç, Telci C, Erkilic EG. The effect of *in vitro* competition on shoot regeneration from hypocotyl explants of *Linum usitatissimum*. Turkish Journal of Botany 2011; 35: 211-218.

[27] Yıldız M, Ulukan H, Özbay A. The effect of different growth medium, gelling agents and explant age on shoot regeneration from hypocotyl explants in flax (*Linum usitatissimum* L.). XIIIrd Biotechnology Congress, Canakkale, Turkey; 2003.

[28] Evers PW. Growth and morphogenesis of shoot initials of Douglas fir, *Pseudotsuga mensiesii* (Mirb.) Franco, *in vitro*. Diss. Agric. Univ. Wageningen, the Netherlands. Article 1-6; 1984

[29] Wilson JB. Shoot competition and root competition. Journal of Applied Ecology 1998; 25: 279-296.

[30] McPhee CS, Aarssen LW. The seperation of above- and below-ground competion in plants. A review and critique of methodology. Plant Ecology 2001; 152: 119-136.

[31] [31] De Klerk GJ. Stress in plants cultured *in vitro*. Propagation of Ornamental Plants 2007; 7(3): 129-137.

[32] Yıldız M. Evaluation of the effect of *in vitro* stress and competition on tissue culture response of flax. Biologia Plantarum 2011; 55(3): 541-544.

[33] Stoffella PJ, Bryan HH. Plant population influences growth and yields of bell pepper. Journal of the American Society for Horticultural Science 1988; 113: 835-839.

[34] Decoteau DR, Grahan HAH. Plant spatial arrangement affects growth, yield and pod distribution of cayenne peppers. HortScience 1994; 29: 149-151.

[35] Jolliffe PA, Gaye MM. Dynamics of growth and yield component responses of bell peppers (Capsicum annuum L.) to row covers and population density. Scientia Horticulturae 1995; 62: 153-164.

[36] Morgade L, Willey R. Effect of plant population and nitrogen fertilizer on yield and efficiency of maize/bean intercropping. *Pesquisa* Agropecuária Brasileira 2003; 38: 1257-1264.

[37] Abubaker S. Effect of plant density on flowering date, yield and quality attribute of bush beans (*Phaseolus vulgaris* L.) under center pivot irrigation system. American Journal of Agricultural and Biological Sciences 2008; 3: 666-668.

[38] Asghari MT, Daneshian J, Farahani AA. Effects of drought stress and planting density on quantity and morphological characteristics of chicory (*Cichorium intybus* L.). Asian Journal of Agricultural Sciences 2009; 1: 12-14.

[39] Gersani M, Brown JS, O'Brien E, Mania GM, Abramsky Z. Tragedy of the commons as a result of root competition. Journal of Ecology 2001; 89: 660-669.

[40] Maina GG, Brown JS, Gersani M. Intra-plant *versus* inter-plant root competiton in beans: avoidance, resource matching or tragedy of the commons. Plant Ecology 2002; 160: 235-247.

[41] Yildiz M, Ozgen M. The effect of a submersion pretreatment on in vitro explant growth and shoot regeneration from hypocotyls of flax (*Linum usitatissimum*). Plant Cell Tissue and Organ Culture 2004; 77: 111-115.

[42] Dale JE. The control of leaf expansion. Annual Review of Plant Physiology 1988; 39: 267-295.

[43] Sunderland N. Cell division and expansion in the growth of the leaf. Journal of Experimental Botany 1960; 11: 68-80.

[44] Ozcan S, Yildiz M, Sancak C, Ozgen M. Adventitious shoot regeneration in sainfoin (*Onobrychis viciifolia* Scop.). Turkish Journal of Botany 1996; 20: 497-501.

[45] Yildiz M, Er C. The effect of sodium hypochlorite solutions on in vitro seedling growth and shoot regeneration of flax (*Linum usitatissimum*). Naturwissenschaften 2002; 89, 259-261.

[46] Allan A. (1991). Plant Cell Culture. In: Stafford A, Warren G. (eds.) Plant Cell and Culture. Open University Press, Milton Keynes; 1991. p1-24.

[47] Dixon RA, Gonzales RA. Plant Cell and Tissue Culture: A Practical Approach, 2nd ed. Oxford University Press, Oxford; 1994.

[48] Yildiz M, Avci M, Ozgen M. Studies on sterilization and medium preparation techniques in sugarbeet (*Beta vulgaris* L.) regeneration. In: Ulrich P. (ed.) Turkish–German Agricultural Research Symposium V, Akdeniz University, Antalya, Turkey; 1997.

[49] Dunn CG. Food Preservatives. In: Lawrence CA, Block SS. (eds.) Disinfection, Sterilization, and Preservation. Lea and Febiger; 1968. p632-651.

[50] Mercer WA, Somers II. Chlorine in food plant sanitation. Advences in Food Research 1957; 7: 129-169.

[51] Smith CR. Mycobactericidal Agents. In: Lawrence CA, Block SS. (eds.) Disinfection, Sterilization, and Preservation. Lea and Febiger; 1968. p504-514.

[52] Spaulding EH. Chemical Disinfection of Medical and Surgical Materials. In: Lawrence CA, Block SS. (eds.) Disinfection, Sterilization, and Preservation. Lea and Febiger; 1968. p517-531.

[53] Bietz JA, Sandford PA Reaction of sodium hypochlorite with amines and amides: Automation of the method. Analytical Biochemistry 1971; 44: 122-133.

[54] Kantouch A, Ardel-Fattah SH. Action of sodium hypochlorite on a-amino acids. Chemicke Zvesti 1971; 25: 222-230.

[55] Hayatsu H, Pan S, Ukita T. Reaction of sodium hypochlorite with nucleic acids and their constituents. Chemical and Pharmaceutical Bulletin 1971; 19: 2189-2192.

[56] Sandford PA, Nafziger AJ, Jeanes A. Reaction of sodium hypochlorite with amines and amides: A new method for quantitating amino sugars in manomeric form. Analytical Biochemistry 1971a; 42: 422-436.

[57] Sandford PA, Nafziger AJ, Jeanes A. Reactions of sodium hypochlorite with amines and amides: A new method for quantitating polysaccharides containing hexosamines. Analytical Biochemistry 1971b; 44: 111-121.

[58] Hsiao AI, Hans JA. Application of sodium hypochlorite seed viability test to wild oat populations with different dormancy characteristics. Canadian Journal of Plant Science 1981; 61, 115-122.

[59] Hsiao AI, Quick AW. Action of sodium hypochlorite and hydrogene peroxide on seed dormancy and germination of wild oats, Avena fatua L. Weed Research 1984; 24, 411-419.

[60] Telci C, Yildiz M, Pelit S, Onol B, Erkilic EG. Kendir, H. The effect of surface-disinfection process on dormancy-breaking, seed germination, and seedling growth of Lathyrus chrysanthus Boiss. under in vitro conditions. Propagation of Ornamental Plants 2011; 11: 10-16.

[61] Racoppi F. Domestic Bleaches Containing Sodium Hypochlorite, Procter and Gamble, Italia SPA, Product Development Department, Rome; 1990.

[62] Schull W. Temperature and rate of moisture intake in seeds. Botanical Gazette 1920; 69: 361-390.

[63] Murashige T, Skoog F. A revised medium for rapid growth and bioassays with tobacco tissue cultures. Physiologia Plantarum 1962; 15: 431-497.

[64] Gamborg OL, Miller RA, Ojima K. Nutrient requirements of suspension cultures of soybean root cells. Experimental Cell Research 1968; 50: 151-158.

[65] Chu CC. The N6 medium and its applications to anther culture of cereal crops. In: Proceedings of Symposium on Plant Tissue Culture. Science Press, Beijing; 1978.

[66] Nitsch JP, Nitsch C. Haploid plants from pollen grains. Science 1969; 163: 85-87.

[67] Driver JA, Kuniyuki AH. In vitro propagation of paradox walnut rootstocks. Hortscience 1984; 19: 507-509.

[68] Lloyd G, McCown B. Commercially feasible micropropagation of mountain laurel, *Kalmia latifolia*, by use of shoot tip culture. *International Plant Propagator's Society Combined Proceedings;* 1980.

[69] Gamborg OL, Phillips GC. Plant Cell Tissue Organ Culture. Narosa Publishing House, New Delhi; 1995.

[70] Fisher RA. Influence of Water Stress on Crop Yield in Semi Arid Regions. In: Turner NC, Kramer P (eds.) Adaptation of Plants to Water and High Temperature Stress. Willey and Son, New York; 1980. P323-340.

[71] Kriedeman PE, Barrs HD. Citrus Orchards. In: Koziowski TT (ed.) Water Deficit and Plant Growth. Academic Press, New York; 1981. p325-417.

[72] Hadas A. Water uptake and germination of Leguminous seeds under changing external water potential in osmotic solutions. Journal of Experimental Botany 1976; 27(3): 480-489.

[73] Hsiao TC. Plant responses to water stress. Annual Review of Plant Physiology 1973; 24: 219-270.

[74] Van Volkenburg E, Boyer JS. Inhibitory effects of water deficit on maize leaf elongation. Plant Physiology 1985; 77: 190-194.

[75] Kahn A. An analysis 'dark-osmotic inhibition' of germination of lettuce seeds. Plant Physiology 1960; 35: 1-7

[76] Morgan PW. Effects of Abiotic Stresses on Plant Hormone Systems. In: Koziowski TT (ed.) Stress Responses in Plants: Adaptation and Acclimation Mechanism. New York: Wiley-Liss; 1990. p113-146.

[77] Yildiz M, Ozcan S, Telci C, Day S, Özat H. The effect of drying and submersion pretreatment on adventitious shoot Regeneration from hypocotyl explants of flax (*Linum usitatissimum* L.). Turkish Journal of Botany 2010; 34: 323-328

[78] Christmann A, Hoffman T, Teplova I, Grill E, Muller A. Generation of active pools for abscisic acid revealed by *in vivo* imaging of water-stressed *Arabidopsis*. Plant Physiology 2005; 137: 209-219.

[79] Okubo H, Wada K, Uemoto S. In vitro morphogenetic response and distribution of endogenous plant hormones in hypocotyl segments of snapdragon (*Antirrhinum majus* L.). Plant Cell Reports 1991; 10: 501-504.

[80] Fatima Z, Mujib A, Fatima S, Arshi A, Umar S. Callus induction, biomass growth, and plant regeneration in *Digitalis lanata* Ehrh.: influence of plant growth regulators and carbonhydrates. Turkish Journal of Botany 2009; 33: 393-405.

[81] Emerson R. Chlorophyll content and the rate of photosynthesis. Proceedings of the National Academy of Sciences of the United States of America 1929; 15(3): 281-284.

[82] Pal RN, Laloraya MM. Effect of calcium levels on chlorophyll synthesis in peanut and linseed plants. Biochemie und Physiologie der Pflanezen 1972; 163: 443-449.

[83] Wright GC, Nageswara RRC, Farquhar GD. Water use efficiency and carbon isotope discrimination in peanut under water deficit conditions. Crop Science 1994; 34: 92-97.

[84] Nageswara RRC, Talwar HS, Wright GC. Rapid assessment of specific leaf area and leaf nitrogen in peanut (*Arachis hypogaea* L.) using chlorophyll meter. Journal of Agronomy and Crop Science 2001; 189: 175-182.

[85] Yang X, Wang X, Wei M. Response of photosynthesis in the leaves of cucumber seedlings to light intensity and CO_2 concentration under nitrate stress. Turkish Journal of Botany 2010; 34: 303-310.

[86] Rensburg LV, Kruger GHJ. Evaluation of components of oxidative stres metabolism for use in selection of drought tolerant cultivars of *Nicotiana tabacum* L. Journal of Plant Physiology 1994; 143: 730-737.

[87] Kyparissis A, Petropoulou Y, Manetas Y. Summer survival of leaves in a soft-leaved shrub (*Phlomis fruticosa* L., Labiatae) under Mediterranean field conditions: Avoidance of photoinhibitory damage through decreased chlorophyll contents. Journal of Experimental Botany 1995; 46: 1825-1831.

[88] Jagtap V, Bhargava S, Sterb P, Feierabend J. Comparative effect of water, heat and light stresses on photosynthetic reactions in *Sorghum bicolor* (L.) Moench. Journal of Experimental Botany 1998; 49: 1715-1721.

[89] Gireesh R. Proximate composition, chlorophyll a, and carotenoid content in *Dunaliella salina* (Dunal) Teod (Chlorophycea: Dunaliellaceae) cultured with cost-effective seaweed liquid fertilizer medium. Turkish Journal of Botany 2009; 33: 21-26.

Chapter 4

In vitro Tissue Culture, a Tool for the Study and Breeding of Plants Subjected to Abiotic Stress Conditions

Rosa Mª Pérez-Clemente and Aurelio Gómez-Cadenas

Additional information is available at the end of the chapter

1. Introduction

Abiotic stress factors are the main limitation to plant growth and yield in agriculture. Among them, drought stress caused by water deficit, is probably the most impacting adverse condition and the most widely encountered by plants, not only in crop fields but also in wild environments. According to published statistics, the percentage of drought-affected land area in the world in 2000 was double that of 1970 [1].

Another major environmental factor that limits crop productivity, mainly in arid and semi-arid regions is high salinity. Approximately 19.5% of the irrigated soils in the world have elevated concentrations of salts either in the soil or in the irrigation water [2], damaging both the economy and the environment [3, 4]. The deleterious effects of salinity on plant growth are associated with low osmotic potential of soil solution (water stress), nutritional imbalance, specific ion effect (salt stress), or a combination of these factors [5].

Abiotic stress leads to a series of morphological, physiological, biochemical, and molecular changes that adversely affect plant growth and productivity [6]. Drought, salinity, extreme temperatures, and oxidative stress are often interconnected, and may induce similar cellular damage (for more details see [7]).

During the course of its evolution, plants have developed mechanisms to cope with and adapt to different types of abiotic and biotic stress. Plants face adverse environmental conditions by regulating specific sets of genes in response to stress signals, which vary depending on factors such as the severity of stress conditions, other environmental factors, and the plant species [8].

The sensing of these stresses induces signaling events that activate ion channels, kinase cascades, production of reactive oxygen species, and accumulation of hormones [9].

These signals ultimately induce expression of specific genes that lead to the assembly of the overall defense reaction. In contrast to plant resistance to biotic stresses, which is mostly dependent on monogenic traits, the genetically complex responses to abiotic stresses are multigenic, and thus more difficult to control and engineer [10].

The conventional breeding programs are being used to integrate genes of interest from inter crossing genera and species into the crops to induce stress tolerance. However, in many cases, these conventional breeding methods have failed to provide desirable results [11].

In recent decades, the use of techniques based on *in vitro* plant tissue culture, has made possible the development of biotechnological tools for addressing the critical problems of crop improvement for sustainable agriculture. Among the available biotechnological tools for crop breeding, genetic engineering based on introgression of genes that are known to be involved in plant stress response and *in vitro* selection through the application of selective pressure in culture conditions, for developing stress tolerant plants, have proved to be the most effective approaches [12].

On the other hand, it is often difficult to analyze the response of plants to different abiotic stresses in the field or in greenhouse conditions, due to complex and variable nature of these stresses. *In vitro* tissue culture-based tools have also allowed a deeper understanding of the physiology and biochemistry in plants cultured under adverse environmental conditions [13].

In this work, the progress made towards the development of abiotic stress-tolerant plants through tissue culture-based approaches is described. The achievements in the better understanding of physiological and biochemical changes in plants under *in vitro* stress conditions are also reviewed.

2. Somaclonal variation

Somaclonal variation is defined as the genetic and phenotypic variation among clonally propagated plants of a single donor clone. It is well known that genetic variations occur in undifferentiated cells, isolated protoplasts, calli, tissues and morphological traits of regenerated plants. The cause of variation is mostly attributed to changes in the chromosome number and structure. Generally, the term somaclonal variation is used for genetic variability present among all kinds of cells/plants obtained from cells cultured *in vitro* [14].

Plants regenerated from tissue and cell cultures show heritable variation for both qualitative and quantitative traits. Somaclonal variation caused by the process of tissue culture is also called tissue culture-induced variation to more specifically define the inducing environment [15]. The occurrence of uncontrolled and spontaneous variation during the culture process is an unexpected and mostly undesired phenomenon when plants are micropropagated at the commercial scale [16]. However, apart from these negative effects, its usefulness in crop breeding through creation of novel variants has been extensively reported [17]. Induced somaclonal variation can be used for genetic manipulation of crops with polygenic traits

[18]. The new varieties derived from *in vitro* tissue culture could exhibit disease resistance and improvement in quality as well as better yield [19].

Somaclonal variants can be detected using various techniques which are broadly categorized as morphological, physiological/biochemical and molecular detection techniques. There are two main approaches for the isolation of somaclonal variants: screening and cell selection.

Screening involves the observation of a large number of cells or regenerated plants for the detection of variant individuals. Mutants for several traits can be far more easily isolated from cell cultures than from whole plant populations. This is because a large number of cells can be easily and effectively screened for mutant traits. Screening of as many plants would be very difficult, ordinarily impossible [17]. Mutants can be effectively selected for disease resistance, improvement of nutritional quality, adaptation to stress conditions, e.g., saline, soils, low temperature, toxic metals, resistance to herbicides and to increase the biosynthesis of plant products used for medicinal or industrial purposes. Screening has been profitably and widely employed for the isolation of cell clones that produce higher quantities of certain biochemicals [20].

In the cell selection approach, a suitable pressure is applied to permit the preferential survival/growth of variant cells. Selection strategies have been successfully developed for the recovery of genotypes resistant to various toxins, herbicides, high salt concentration etc. [21]. When the selection pressure allows only the mutant cells to survive or divide, it is called positive selection. On the other hand, in the case of negative selection, the wild type cells divide normally and therefore are killed by a counter selection agent, e.g., 5-Bromodeoxyuridine, or arsenate. The mutant cells are unable to divide as a result of which they escape the counter selection agent. These cells are subsequently rescued by removal of the counter selection agent [11].

3. *In vitro* selection of plants tolerant to abiotic stress

Many studies have reported that the *in vitro* culture alone or combined with mutagenesis, induced with physicochemical or biological agents, can be exploited to increase genetic variability and mutants, as a potential source of new commercial cultivars [22]. *In vitro* culture environments can be mutagenic and plants regenerated from organ cultures, calli, protoplasts and via somatic embryogenesis sometimes exhibit phenotypic and/or genotypic variations [22].

It is important to point that tissue culture increases the efficiency of mutagenic treatments and allows handling of large populations and rapid cloning of selected variants [17]. The similarities of the effects induced by the stress in the plant cultured *in vitro* and *in vivo* conditions suggest that the *in vitro* system can be used as an alternative to field evaluations for studying the general effect of water-stress on plant growth and development.

The most widely used method for the selection of genotypes tolerant to abiotic stress is the *in vitro* selection pressure technique. This is based on the *in vitro* culture of plant cells, tissues or organs on a medium supplemented with selective agents, allowing selecting and regenerating plants with desirable characteristics. In table 1 a list of species in which this technique has been successfully applied to obtain genotypes with increased resistance to different abiotic stresses is shown.

Plant species	stress	References
Chrysanthemum morifolium (chrysanthemum)	salt	[66]
Brassica napus (rapeseed)	salt	[67]
Citrus aurantium (sour orange)	salt	[68]
Lycopersicon esculentum (tomato)	salt	[69]
Dendrocalamus strictus (bamboo)	salt	[11]
Ipomoea batatas (sweet potato)	salt	[70]
Saccharum sp. (sugarcane)	salt	[71]
Solanum tuberosum (potato)	salt	[32], [72]
Triticum aestivum (wheat)	salt	[21], [73]
Arachis hypogaea (groundnut)	drought	[74]
Brassica juncea (indian mustard)	drought	[75]
Prunus avium (colt cherry)	drought	[72]
Saccharum sp. (sugarcane)	drought	[76]
Oryza sativa (rice)	drought/chilling/Al	[77]
Triticum aestivum (wheat)	drought/frost	[73]
Glycine max (soybean)	Al	[78]
Setaria italica (millet)	Zn	[79]

Table 1. *In vitro* selection for increased resistance to abiotic stresses.

The most important successes on this respect are described below:

3.1. *In vitro* selection of salt-tolerant plants

The problem of soil salinity has been aggravated during the last decades as a consequence of some agricultural practices such as irrigation and poor drainage systems. As described in the introduction, it has been estimated that around 20 % of the irrigated land in the world is affected by salinity, and it is expected that the increase of salinization in agricultural fields will reduce the land available for cultivation by 30% in the next 25 years and up to 50% by the year 2050 [23].

The *in vitro* selection pressure technique has been effectively utilized to induce tolerance to salt stress in plants through the use of salts as a selective agent, allowing the preferential survival and growth of desired genotypes. This approach has being done using a number of plant materials (callus, suspension cultures, somatic embryos, shoot cultures, etc.) which has been screened for variation in their ability to tolerate relatively high levels of salt in the culture media. In most of the studies, the salt used has been NaCl [24].

Several researchers have compared the response of other Cl⁻ and SO_4^{2-} salts including KCl, Na_2SO_4, and $MgSO_4$ during *in vitro* screening. This use of multiple salts as a selection pressure parallels the salinity under field conditions and may be a better choice [11].

3.2. *In vitro* selection of drought-tolerant plants

Drought is a major abiotic stress which causes important agricultural losses, mainly in arid and semiarid areas. Drought stress causes moisture depletion in soil and water deficit with a decrease of water potential in plant tissues. *In vitro* culture has been used to obtain drought-tolerant plants assuming that there is a correlation between cellular and *in vivo* plant responses [25]. During the last years, *in vitro* selection for cells exhibiting increased tolerance to water or drought stress has been reported (Table 1).

Polyethylene glycol (PEG), sucrose, mannitol or sorbitol have been used by several workers as osmotic stress agents for *in vitro* selection [25; 26] However, PEG has been the most extensively used to stimulate water stress in plants. This compound of high molecular weight is a non-penetrating inert osmoticum that reduces water potential of nutrient solutions without being taken up by the plant or being phytotoxic [26]. Because PEG does not enter the apoplast, water is withdrawn not only from the cell but also from the cell wall. Therefore, PEG solutions mimic dry soil more closely than solutions of compounds with low molecular weights, which infiltrate the cell wall with solute [27].

Besides salt and drought, a few reports are also available for the development of plants tolerant to other abiotic stress (metal, chilling, UV and frost) through *in vitro* selection (Reviewed in [11]).

4. Characterization of salt- or drought-tolerant plants during *in vitro* selection

A second step in the process of obtaining genotypes more tolerant to a particular stress condition is the characterization of the regenerants that survive to the imposed pressure selection under *in vitro* conditions. Salinity and drought affect many physiological processes such as reductions of cell growth, leaf area, biomass and yield. The activation of the plant antioxidant defense system has been positively associated with salt and drought tolerance [11], and the same pattern has been confirmed on *in vitro* cultures [28]. Therefore, by measuring antioxidant activities *in vitro*, a rapid preliminary selection of tolerant genotypes could be performed. In fact, different authors have determined the main antioxidant enzymes such as superoxide dismutase (SOD), ascorbate peroxidase (APX), catalase (CAT) and glutathione reductase (GR) [29].

Lipids play an important role as the structural constituent of most of the cellular membranes [30]. Moreover, often there is no need of intact plants to perform the initial selection, as in the case of callus culture that can be used as a plant material for the selection of tolerant genotypes. As an example it is well known that free radical-induced peroxidation of lipid membrane is a sign of stress-induced damage at cellular level. Therefore, the level of malonyldialdehyde, produced during peroxidation of membrane lipids, is often used as an indicator of oxidative damage [31]. It has been reported that selected callus lines of *Solanum tuberosum* subjected to NaCl showed an increase in lipid peroxidation in comparison with salt tolerant lines [32].

For overcoming salt or drought stress, plants have developed protective mechanisms including osmotic adjustment that is usually accomplished by accumulation of compatible solutes such as proline, glycine betaine and polyols [33]. It has been also reported that proline levels increased in response to water stress in tomato calli [34].Taking into account the generated knowledge about plants responses to abiotic stress conditions, the determination of antioxidant enzyme activities, and levels of malonyldialdehyde and proline in plants recovered under selective conditions may help to isolated the most tolerant genotypes.

In recent years, both basic and applied research has led to understand the mechanisms underlying the stress response and the identification of the specific genes/metabolites that are responsible for tolerance phenotypes the "omics" approaches have had a significant development. Through the application of transcriptomics hundreds of genes have been linked with environmental stress responses and regulatory networks of gene expression have been delineated [35]. Moreover, plant tolerance to abiotic stress conditions has been associated with changes in proteome composition. Since proteins are directly involved in plant stress response, proteomics studies can significantly contribute to unravel the possible relationships between protein abundance and plant stress acclimation [36].

Relatively less is known about changes at the metabolomic level. Metabolome analysis has become a valuable tool to study plant metabolic changes that occur in response to abiotic stresses. This approach has already enabled to identify different compounds whose accumulation is affected by exposure to stress conditions. However, much work is still required to identify novel metabolites and pathways not yet linked to stress response and tolerance [37].

In this context, an integrated approach incorporating *in vitro* plant tissue culture to proteomics and metabolomics technique, can contribute to elucidate the metabolites involved in stress response and select desired genotypes at early stages of plant development or at callus stage [38].

5. Transgenesis for abiotic stress tolerance

Transgenic approaches are among the available tools for plant improvement programs based on biotechnological methodologies. Nowadays, many mechanisms and gene families, which confer improved productivity and adaptation to abiotic stresses, are known. These gene families can be manipulated into novel combinations, expressed ectopically, or transferred to species in which they do not naturally occur. Therefore, the possibility to transform the major crop species with genes from any biological source (plant, animal, microbial) is an extremely powerful tool for molecular plant breeding [39].

To date, successes in genetic improvement of environmental stress resistance have involved manipulation of a single or a few genes involved in signaling/regulatory pathways or that encode enzymes involved in these pathways (such as osmolytes/compatible solutes, antioxidants, molecular chaperones/osmoprotectants, and water and ion transporters [8]. The disadvantage of this approach is that there are numerous interacting genes involved, and efforts to improve crop drought tolerance through manipulation of one or a few of them is often associated with other, often undesirable, pleiotropic and phenotypic alterations [8].

The plant hormone abscisic acid (ABA) regulates the adaptive response of plants to environmental stresses such as drought, salinity, and chilling via diverse physiological and developmental processes [40, 41].

The ABA biosynthetic pathway has been deeply studied and many of the key enzymes involved in ABA synthesis have been used in transgenic plants in relation to improving abiotic stress tolerance [42]. Transgenic plants overexpressing the genes involved in ABA synthesis showed increased tolerance to drought and salinity stress [42, 43]. Similarly, many studies have illustrated the potential of manipulating CBF/DREB genes to confer improved drought tolerance [44, 45].

Another mechanism involved in plant protection to osmotic stress associated to drought and salinity involves the upregulation of compatible solutes that function primarily to maintain cell turgor, but are also involved in avoiding oxidative damage and chaperoning through direct stabilization of membranes and/or proteins [46; 47]. Many genes involved in the synthesis of these osmoprotectants have been explored for their potential in engineering plant abiotic stress tolerance [10, 47].

The cellular and metabolic processes involved in salt stress are similar to those occurring in drought-affected plants and are responses to the osmotic effect of salt [48, 49]. As described above, the use of genes related to osmoprotectant synthesis has been successfully used in developing drought-tolerant crops and the transfer of glycine betaine intermediates have improved the drought and salt tolerance of transgenic plants in many cases [50].

The amino acid proline is known to occur widely in higher plants and normally accumulates in large quantities in response to environmental stresses [51, 52]. The osmoprotectant role of proline has been verified in some crops by overexpressing genes involved in proline synthesis [53].

Other approaches successfully developed in a variety of crops to obtain abiotic-stress-tolerant plants by transgenesis, have been manipulation of transcription factors (TFs), late embryogenesis abundant (LEA) proteins, and antioxidant proteins [54].On the other hand, the use of genetic and genomic analysis to identify DNA molecular markers associated to stress resistance can facilitate breeding strategies for crop improvement. This approach is particularly useful when targeting characters controlled by several genes, as in the case of most abiotic stress.

The potential to map different Quantitative Trait Loci (QTL) contributing to an agronomical trait and to identify linked molecular markers opens up the possibility to transfer simultaneously several QTLs and to pyramid QTLs for several agronomical traits in one improved cultivar [55]. However, the application of molecular markers in breeding programs requires preliminary studies to identify and validate potential markers [55].

Although the use of Marker-Assisted Selection may be helpful for crop improvement, its practical application in genetic improvement of resistance or tolerance to stress has been limited since no many stress tolerance QTL have been identified [56]. For future biotechnology improvements such as tolerance to drought or nutrient limitation, forward breeding will be necessary to co-optimize transgenic expression and genetic background because endogenous genes and environmental factors may have the potential to influence the phenotypes resulting from transgenic modifications [57].

It is important to point that genetic modification of higher plants by introducing DNA into their cells is a highly complex process. Practically any plant transformation experiment relies at some point on cell and tissue culture. Although the development transformation methods that avoid plant tissue culture have been described for *Arabidopsis*, and have been extended to a few crops, the ability to regenerate plants from isolated cells or tissues *in vitro* is needed for most plant transformation systems. Not all plant tissue is suited to every plant

Plant species	stress	References
Ipomoea batatas (sweet potato)	salt	[16]
Triticum aestivum (wheat)	salt	[80]
Oryza sativa (rice)	salt/drought	[81]
Lactuca sativa (lettuce)	drought/cold	[82]
Brassica juncea (mustard)	Cd	[83]
Medicago sativa (alfalfa)	freezing	[84]
Gossypium hirsutum (cotton)	chilling	[85]
Solanum tuberosum (potato)	salt-drought	[86]
Avena sativa (oat)	osmotic	[87]
Daucus carota (carrot)	salt	[88]
Zea mays (maize)	drought	[89]
Pinus taeda (loblolly pine)	salt	[90]
Populus tomentosa (chinese white poplar)	salt	[91]
Citrus sinensis x Poncirus trifoliate (Carrizo citrange)	drought	[92]
Petunia sp. (petunia)	drought	[93]
Nicotiana tabacum (tobacco)	freezing/Al	[94]
Brassica juncea (indian mustard)	As and Cd	[95]
Brassica napus (rape)	freezing	[96]
Pinus virginiana (Virginia pine)	metal	[90]
Hordeum vulgare (barley)	Al	[97]
Alyssum sp. (alyssum)	Ni	[98]

Table 2. Genetic transformation for increased resistance to abiotic stresses.

transformation method, and not all plant species can be regenerated by every method [58]. There is, therefore, a need to find both a suitable plant tissue culture/regeneration regime and a compatible plant transformation methodology. Today, many agronomical and horticultural important species are routinely transformed, and the list of species that is susceptible to *Agrobacterium*-mediated transformation seems to grow daily (Table 2).

6. *In vitro* tissue culture as a tool for physiological and biochemical studies in plants

Because of the great interest for both basic and applied research, many scientific endeavours have long addressed the understanding of the mechanisms underlying the stress response and the identification of the specific genes/metabolites that are responsible for tolerance phenotypes.

In the last decades, *in vitro* culture of plants has become an integral part of advances in plant science research. Plant tissue culture techniques allow for close monitoring and precise manipulation of plant growth and development, indeed, the *in vitro* system offers the advantage that relatively little space is needed to culture plants and this system allows a rigorous control of physical environment and nutrient status parameters, which are difficult to regulate with traditional experimental system [59]. Furthermore, any complex organ-organ and plant-environment interaction can be controlled or removed, and the level of stress can be accurately and conveniently controlled [60]. All this together makes that some aspects of plant growth, that were barely understood before the advancement of the science of tissue culture, such as the metabolism and interaction of plant hormones, as well as their physiological effects can be deeply studied [61].

Shoot apex culture has been widely used to evaluate plant physiological responses to salinity and osmotic stress in various species, including apple [59], olive [62] and tomato [63]. With regard to the whole plant, a similar response to salt stress could be expected in plantlets grown through *in vitro* shoot apex culture [63], because such explants can be considered mini-replicas of a plant with anatomical organization and ability to root and grow into whole plant.

We have previously described the use of an *in vitro* tissue culture technique to study the performance of different citrus genotypes cultured under salt stress conditions, avoiding the effect of the root by culturing shoots without the root system. The method proved to be a good tool for studying biochemical processes involved in the response of citrus to salt stress [64]. Some citrus genotypes have been classified as relatively salt tolerant under field conditions due to their ability to restrict chloride ions to roots while others have proved to be more sensitive to salinity [65].

In vitro tissue culture approach allowed us to observe that when shoots are cultured without a root system, all genotypes accumulated the same chloride levels and exhibited similar leaf damage as a consequence of salt stress treatment. There was no increase of malonyldialdehyde levels in any genotype, and common patterns of hormonal signaling were observed among genotypes. On the view of these results we concluded that under the same salt conditions and with the same level of leaf chloride intoxication, no biochemical differences exist among tolerant and sensitive genotypes. This points to the roots as a key organ not only as a filter of chloride ions but also as a signalling system in citrus [64]. *In vitro* tissue culture provided the tools to perform this studies that it would be impossible to carry out with whole plants grown under field or greenhouse conditions.

7. Conclusion

Use of *in vitro* cell and tissue-based systems offers a remarkable tool for dissecting the physiological, biochemical and molecular regulation of plant development and stress response phenomena. In recent years, considerable progress has been made regarding the development and isolation of stress tolerant genotypes by using *in vitro* techniques.

The most successful applied tools have been the induction of somaclonal variation and *in vitro* selection of plants tolerant to different abiotic stresses and the development of transgenic genotypes throughout different approaches.

In vitro selection makes possible to save the time required for developing disease resistant and abiotic stress tolerant lines of commercial crops and other plant species. However, *in vitro* selected variants should be finally field-tested to confirm the genetic stability of the selected traits under field conditions. The development of *in vitro* selection technology, together with molecular approaches and functional genomics will provide a new opportunity to improve stress tolerance in plants relevant to food production and environmental sustainability.

Development of transgenic plants using biotechnological tools has become important in plant-stress biology. Previous works on genetics and molecular approaches have shown that most of the abiotic stress tolerant traits are multigenic. Therefore, to improve stress tolerance several stress related genes need to be transfered. More recently manipulation of single transcription factors has provide the same effect as manipulation of multiple genes. This has become a promising approach to get abiotic stress tolerant crops.

A limiting factor for the widespread application of this technology is that, with few exceptions, genetic transformation protocols require plant regeneration of transformants using *in vitro* plant tissue culture tools. Although the list of species that are susceptible to *Agrobacterium*-mediated transformation has been increased recently, still are many genotypes for which regeneration protocols are not available.

On the other hand, plant tissue culture is also an invaluable laboratory tool to study basic aspects of plant growth and development, and to manipulate these processes since it makes possible to have a large number of plants in a small space, without the interference of other biotic or abiotic stress factors. It also allows growing plants in the same nutritional and environmental conditions all year around.

Author details

Rosa Mª Pérez-Clemente and Aurelio Gómez-Cadenas
Department of Agricultural Sciences. Universidad Jaume I. Campus Riu Sec. Castellón, Spain

Acknowledgement

This work was supported by the Spanish Ministerio de Economía y competitividad (MINECO) and Universitat Jaume I/Fundació Bancaixa through grants No. AGL2010-22195-C03-01/AGR and P11B2009-01, respectively.

8. References

[1] Isendahl N., Schmidt G. Drought in the Mediterranean-WWF policy proposals; A. WWF Report, Madrid 2006.

[2] Jin T.C., Chang Q., Li W.F., Yin D.X., Li Z.J., Wang D.L., Liu B., Liu L.X. Stress-inducible expression of *GmDREB1* conferred salt tolerance in transgenic alfalfa. Plant Cell Tissue and Organ Culture 2010; 100 219–227.

[3] Rengasamy P. Soil processes affecting crop production in salt-affected soils. Functional Plant Biology 2010; 37 255–263.

[4] Yang Y.L., Shi R.X., Wei X.L., Fan Q., An L.Z. Effect of salinity on antioxidant enzymes in calli of the halophyte *Nitraria tangutorum* Bobr. Plant Cell Tissue and Organ Culture 2010; 102 387–395.

[5] Gómez-Cadenas A., Arbona V., Jacas J., Primo-Millo E., Talon M. Abscisic acid reduces leaf abscission and increases salt tolerance in citrus plants. Journal of Plant Growth Regulation 2003; 21 234- 240.

[6] Wang W.X., Vinocur B., Shoseyov O., Altman A. Biotechnology of plant osmotic stress tolerance: physiological and molecular considerations. Acta Horticulturae 2001; 560 285–292.

[7] Jewell M.C., Campbell B.C, Godwin I.D. transgenic plants for abiotic stress resistance. In: Transgenic Crop Plants. C. Kole et al. (eds.), pringer-Verlag Berlin Heidelberg. 2010

[8] Wang W., Vinocur B., Altman A. Plant responses to drought, salinity and extreme temperatures: towards genetic engineering for stress tolerance. Planta 2003; 218 1–14.

[9] Cheong Y.H., Chang H.S., Gupta R., Wang X., Zhu T., Luan S. Transcriptional profiling reveals novel interactions between wounding, pathogen, abiotic stress, and hormonal responses in Arabidopsis. Plant Physiology 2002; 129: 661–677.

[10] Vinocur B., Altman A. Recent advances in engineering plant tolerance to abiotic stress: achievements and limitations. Current Opinion in Biotechnology 2005; 16 123-132.

[11] Rai M.K., Kalia R .K., Singh R., Gangola M.P., Dhawan A.K. Developing stress tolerant plants through in vitro selection—An overview of the recent progress. Environmental and Experimental Botany 2011; 71 89-98.

[12] Sakhanokho H.F., Kelley R.Y. Influence of salicylic acid on *in vitro* propagation and salt tolerance in *Hibiscus acetosella* and *Hibiscus moscheutos* (cv 'Luna Red') African Journal of Biotechnology 2009; 8 1474–1481.

[13] Benderradji L., Brini F., Kellou K., Ykhelf N., Djekoun A., Masmoudi K., Bouzerour. H. Callus induction, proliferation, and plantlets regeneration of two bread wheat (*Triticum aestivum* L.) genotypes under saline and heat stress conditions. 2012 ISRN Agronomy, Article ID 367851.

[14] Lestari E.G. *In vitro* selection and somaclonal variation for biotic and abiotic stress tolerance. Biodiversitas 2006; 7 297-301.

[15] Kaeppler S.M., Kaepler H.F., Rhee Y. Epigenetic aspects of somaclonal variation in plants. Plant Molecular Biology. 2000; 43: 179-188.

[16] Gao X., Yang D., Cao D., Ao M., Sui X., Wang Q., Kimatu J., Wang L. *In vitro* micropropagation of Freesia hybrida and the assessment of genetic and epigenetic stability in regenerated plantlets. Journal of Plant Growth Regulation 2010; 29 257–267.

[17] Predieri S. Mutation induction and tissue culture in improving fruits. Plant Cell Tissue and Organ Culture 2001; 64 185–210.

[18] Jain S.M. Tissue culture-derived variation in crop improvement. Euphytica 2001; 118 153–166.

[19] Biswas M.K., Dutt M., Roy U.K., Islam R., Hossain M. Development and evaluation of *in vitro* somaclonal variation in strawberry for improved horticultural traits. Scientia Horticulturae 2009; 122 409–416.

[20] Matkowski A. Plant *in vitro* culture for the production of antioxidants — A review. Biotechnology Advances 2008; 26 584-560.

[21] Zair I., Chlyah A., Sabounji K., Tittahsen M., Chlyah H. Salt tolerance improvement in some wheat cultivars after application of *in vitro* selection pressure. Plant Cell Tissue and Organ Culture 2003; 73 237-244.

[22] Orbović V., Ćalović M., Viloria Z., Nielsen B., Gmitter F., Castle W., Grosser J. Analysis of genetic variability in various tissue culture-derived lemon plant populations using RAPD and flow cytometry. Euphytican 2008; 161 329–335.

[23] Rozema J., Flowers T. Crops for a salinized world. Science 2008; 322 1478-1480.

[24] Woodward A.J., Bennett I.J. The effect of salt stressand abscisic acidon proline production, chlorophyll content and growth of *in vitro* propagated shoots of *Eucalyptus camaldulensis*. Plant Cell Tissue and Organ Culture 2005; 82 189-200.

[25] Mohamed M.A.H., Harris, P.J.C., Henderson, J. *In vitro* selection and charac-terisation of a drought tolerant clone of *Tagetes minuta*. Plant Science 2000; 159 213-222.

[26] Hassan N.M., Serag M.S., El-Feky F.M. Changes in nitrogen content and protein profiles following *in vitro* selection of NaCl resistant mung bean and tomato. Acta Physiologiae Plantarum 2004; 26 165-175.

[27] Verslues P.E., Ober E.S., Sharp R.E. Root growth and oxygen relations at low water potentials. Impact of oxygen availability in polyethylene glycol solutions. Plant Physiology 1998; 116 1403–1412.

[28] Lascano H.R. Antioxidant system response of different wheat cultivars under drought: field and *in vitro* studies. Australian Journal of Plant Physiology 2001; 28 1095 – 1102.

[29] Hossain Z., Mandal A.K.A., Datta S.K., Biswas A.K. Development of NaCl tolerant line in *Chrysanthemum morifolium* Ramat. through shoot organogenesis of selected callus line. Journal of Biotechnology. 2007; 129 658-667.

[30] Parida A.K., Das A.B. Salt tolerance and salinity effects on plants: a review. Ecotoxicology and Environmental Safety 2005; 60 324-349.

[31] Demiral T., Turkan, I. Comparative lipid peroxidation, antioxidant defense systems and proline content in roots of two rice cultivars differing in salt tolerance. Environmental and Experimental Botany. 2005; 53 247-257.

[32] Queiros F., Fidalgo F., Santos I., Salema R. *In vitro* selection of salt tolerant cell lines in *Solanum tuberosum* L. Biologia Plantarum. 2007; 51 728-734.

[33] Ghoulam C., Ahmed F., Khalid F. Effects of salt stress on growth, inorganic ions and proline accumulation in relation to osmotic adjustment in five sugar beet cultivars. Environmental and Experimental Botany 2001; 47 139-150.

[34] Aazami M.A., Torabi M., Shekari F. Response of some tomato cultivars to sodium chloride stress under *in vitro* culture condition African Journal of Agricultural Research. 2010; 5 2589 – 2592.

[35] Todaka D., Nakashima K., Shinozaki K., Yamaguchi-Shinozaki K. Toward understanding transcriptional regulatory networks in abiotic stress responses and tolerance in rice. Todaka et al. Rice 2012, 5:6 http://www.thericejournal.com/content/5/1/6.

[36] Kosová, K., Vítámvás, P., Prášil, I.T. Renaut, J. Plant proteome changes under abiotic stress - Contribution of proteomics studies to understanding plant stress response. Journal of Proteomics, 2011; 15 51-58.

[37] Cevallos-Cevallos, J. M., Reyes De Corcuera, J. I., Etxeberria, E., Danyluk, M. D. Rodrick, J. E. Metabolomic analysis in food science: a review. Trends in Food Science and Technology 2009; 20 557-566.

[38] Palama T., Menard P., Fock I., Bourdon E., Govinden-Soulange J., Bahut M., Payet B., Verpoorte R., Kodja H. Shoot differentiation from protocorm callus cultures of *Vanilla planifolia* (Orchidaceae): proteomic and metabolic responses at early stage. BMCP Plant Biology 2010; 10 82.

[39] Miflin B. Crop improvement in the 21st century. Journal of Experimental Botany 2000; 51 1-8.

[40] Gómez-Cadenas A, Tadeo Fr, Primo-Millo E, Talón M. Involvement of abscisic acid and ethylene in the response of citrus seedlings to salt shock. Physiologia Plantarum 1998; 103 475- 484.

[41] Arbona V, Gómez-Cadenas A. Hormonal modulation of citrus responses to flooding. Journal of Plant Growth Regulation 2008; 27 241-250.

[42] Schwartz S.H., Qin X., Zeevaart J.A.D. Elucidation of the indirect pathway of abscisic acid biosynthesis by mutants, genes, and enzymes. Plant Physiology 2003; 131 1591–1601.

[43] Park H.Y., Seok H.Y., Park B.K., Kim S.H., Goh C.H., Bh L., Lee C.H., Moon Y.H. Overexpression of Arabidopsis ZEP enhances tolerance to osmotic stress. Biochemical and Biophysical Research Communications 2008; 375: 80–85.

[44] Al-Abed D., Madasamy P., Talla R., Goldman S., Rudrabhatla S. Genetic engineering of maize with the Arabidopsis DREB1A/CBF3 gene using split-seed explants. Crop Science 2007; 47 2390–2402.

[45] Trujillo L.E., Sotolongo M., Menendez C., Ochogavia M.E., Coll Y., Hernandez I., Borras-Hidalgo O.,Thomma B.P.H.J., Vera P., Hernandez L. SodERF3, a novel sugarcane ethylene responsive factor (ERF), enhances salt and drought tolerance when overexpressed in tobacco plants. Plant and Cell Physiology 2008; 49 512–525.

[46] McNeil S.D., Nuccio M.L., Hanson A.D. Betaines and related osmoprotectants. Targets for metabolic engineering of stress resistance. Plant Physiology 1999; 120 945–949.

[47] Zhang Y.Y., Li Y., Gao T., Zhu H., Wang D.J., Zhang H.W., Ning Y.S., Liu L.J., Wu Y.R., Chu C.C., Guo H.S., Xie Q. 2008 Arabidopsis SDIR1 enhances drought tolerance in crop plants. Biosci Biotechnol Biochem 72: 2251–2254.

[48] Yeo A.R., Lee K.S., Izard P., Boursier P.J., Flowers T.J. Short- and long-term effects of salinity on leaf growth in rice (*Oryza sativa* L.). Journal of Experimental Botany 1991; 42 881–889.

[49] Munns R., Tester M. Mechanisms of salinity tolerance. Annual Review of Plant Physiology 2008; 59 651–681.

[50] Wu W., Su Q., Xia X., Wang Y., Luan Y., An L. The *Suaeda liaotungensis* kitag betaine aldehyde dehydrogenase gene improves salt tolerance of transgenic maize mediated with minimum linear length of DNA fragment. Euphytica 2008; 159 17–25.

[51] Kavi Kishore P.B., Sangam S., Amrutha R.N., Laxmi P.S., Naidu K.R., Rao K.R.S.S., Rao S., Reddy K.J., Theriappan P., Sreenivasulu N. Regulation of proline biosynthesis, degradation, uptake and transport in higher plants: its implications in plant growth and abiotic stress tolerance. Current Science 2005; 88 424–438.

[52] Ashraf M., Foolad M.R. Roles of glycine betaine and proline in improving plant abiotic stress resistance. Environmental and Experimental Botany 2007; 59 206–216.

[53] Hmida-Sayari A., Gargouri-Bouzid R., Bidani A., Jaoua L., Savoure A., Jaoua S. Overexpression of D^1-pyrroline-5-carboxylate synthetase increases proline production and confers salt tolerance in transgenic potato plants. Plant Science 2005; 169 746–752.

[54] Umezawa T., Fujita M., Fujita Y., Yamaguchi-Shinozaki K., Shinozaki K., Engineering drought tolerance in plants: discovering and tailoring genes to unlock the future. Current Opinion in Biotechnology 2006; 17 113-122.

[55] Yu K.F., Park S.J., Zhang B.L., Haffner M. Poysa V. An SSR marker in the nitrate reductase gene of common bean is tightly linked to a major gene conferring resistance to common bacterial blight. Euphytica 2004; 138: 89–95.

[56] Foolad, M.R., Recent advances in genetics of salt tolerance tomato. Plant Cell Tissue and Organ Culture 2004; 76: 101–119.

[57] Mumm R.H. Backcross versus forward breeding in the development of transgenic maize hybrids: theory and practice. Crop Science *(Suppl 3)* 2007; 47: 164–171.

[58] Benson E.E. Special symposium: *in vitro* plant recalcitrance *In vitro* plant recalcitrance: an introduction. *In Vitro* Cellular and Developmental Biology Plant 2000; 36:141–148.

[59] Shibli R.A., Smith M.A.L., Spomer L.A. Osmotic adjustment and growth responses of three *Chrysanthemum morifolium* Ramat cultivars to osmotic stress induced *in vitro*. Journal of Plant Nutrition, 1992; 15 1374-1381.

[60] Stephen R.G., Zeng L., Shannon M.C., Roberts S.R. Rice is more sensitive to salinity than previously thought. Annual Report of U.S. Department of Agriculture's Research Service, pp: 481. 2002.

[61] Bairu M.W., Kane M.E. Physiological and developmental problems encountered by *in vitro* cultured plants. Plant Growth Regulation 2011; 63 101–103

[62] Shibli R.A., Al-Juboory K. Comparative response of 'Nabali' olive microshoot, callus and suspension cell cultures to salinity and water deficit. Journal of Plant Nutrition 2002; 25 61-74.

[63] Cano E.A., Perez-Alfocea F., Moreno V., Caro M., Bolarin, M.C. Evaluation of salt tolerance in cultivated and wild tomato species through *in vitro* shoot apex culture. Plant Cell, Tissue and Organ Culture 1998; 53 19-26.

[64] Montoliu A., López-Climent M.F., Arbona V., Pérez-Clemente R.M., Gómez-Cadenas A. A novel *in vitro* tissue culture approach to study salt stress responses in citrus. Plant Growth Regulation 2009; 59 179-187.

[65] López-Climent M.F., Arbona V., Pérez-Clemente R.M., Gómez-Cadenas A. Relationship between salt tolerance and photosynthetic machinery performance in citrus. Environmental and Experimental Botany 2008; 62 176–184.

[66] Hossain Z., Mandal A.K.A., Datta S.K., Biswas A.K., Development of NaCl tolerant line in *Chrysanthemum morifolium* Ramat. through shoot organogenesis of selected callus line. Journal of Biotechnology. 2007; 129 658-667.

[67] Rahman, M.H., Krishnaraj, S., Thorpe, T.A. Selection for salt tolerance *in vitro* using microspore-derived embryos of *Brassica napus* cv. Topas, and the characterization of putative tolerant plants. In Vitro Cellular and Developmental Biology - Plant 1995; 31 116-121.

[68] Koc N.K., Bas B., Koc M., Kusek M. Investigations of *in vitro* selection for salt tolerant lines in sour orange (*Citrus aurantium* L.). Biotechnology 2009; 8 155-159.

[69] Kripkyy O., Kerkeb L., Molina A., Belver A., Rodrigues Rosales P., Donaire P.J., Effects of salt-adaptation and salt-stress on extracellular acidification and microsome phosphohydrolase activities in tomato cell suspensions. Plant Cell Tissue and Organ Culture. 2001; 66 41-47.

[70] He S., Han Y., Wang Y., Zhai H., Liu Q. *In vitro* selection and identification of sweetpotato (*Ipomoea batatas* (L.) Lam.) plants tolerant to NaCl. Plant Cell Tissue and Organ Culture 2009; 96: 69-74.

[71] Gandonou C.B., Errabii T., Abrini J., Idaomar M., Senhaji N.S. Selection of callus cultures of sugarcane (*Saccharum* sp.) tolerant to NaCl and their response to salt stress. Plant Cell Tissue and Organ Culture. 2006; 87 9-16.

[72] Ochatt S.J., Marconi P.L., Radice S., Arnozis P.A., Caso O.H. *In vitro* recurrent selection of potato: production and characterization of salt tolerant cell lines and plants. Plant Cell Tissue and Organ Culture 1999; 55 1-8.

[73] Barakat, M.N., Abdel-Latif, T.H. *In vitro* selection of wheat callus tolerant to high levels of salt and plant regeneration. Euphytica 1996; 91 127-140.

[74] Purushotham M.G., Patil V., Raddey P.C., Prasad T.G., Vajranabhaiah S.N. Development of *in vitro* PEG stress tolerant cell lines in two groundnut (*Arachis hypogaea* L.) genotypes. Indian Journal of Plant Physiology. 1998; 3 49-51.

[75] Gangopadhyay G., Basu S., Gupta S. *In vitro* selection and physiological characterization of NaCl- and mannitol-adapted callus lines in *Brassica juncea*. Plant Cell Tissue and Organ Culture. 1997; 50 161-169.

[76] Errabii T., Gandonou C.B., Essalmani H., Abrini J., Idomar M., Senhaji N.S. Growth, proline and ion accumulation in sugarcane callus cultures under drought-induced osmotic stress and its subsequent relief. African Journal of Biotechnology. 2006; 5 1488-1493.

[77] Roy B., Mandal A.B. Towards development of Al-toxicity tolerant lines in indica rice by exploiting somaclonal variation. Euphytica 2005; 145 221-227.

[78] Mariska I. Peningkatan Ketahanan terhadap Alumunium pada Pertanaan Kedelai melalui Kultur in Vitro. Laporan RUT VIII. Bidang Teknologi Pertanian. Bogor: Balai Penelitian Bioteknologi Pertanian & Kementerian Riset dan Teknologi RI-LIPI. 2003

[79] Samantaray S., Rout G.R., Das P. *In vitro* selection and regeneration of zinc tolerant calli from *Setaria italica* L Plant Science. 1999; 31 201-209.

[80] Laurie S., Feeney K.A., Mathuis F.J., Heard P.J., Brown S.J., Leigh R.A. A role for HKT1 in sodium uptake by wheat roots, Plant Journal 2002; 32 139–49.

[81] Liu K., Lei W., Xu Y., Chen N., Ma Q., Li F., Chong K. Overexpression of OsCOIN, a putative cold inducible zinc finger protein, increased tolerance to chilling,salt and drought, and enhanced praline level in rice. Planta, 2007; 226 1007-1016.

[82] Vanjildorj E., Bae T.W., Riu K.Z., Kim S.Y., Lee H.Y. Overexpression of Arabidopsis ABF3 gene enhances tolerance to drought and cold transgenic lettuce (*Lactuca sativa*). Plant Cell, Tissue and Organ Culture. 2005; 83 41-50.

[83] Zhu Y.L., Pilon-Smits E.A.H., Jouanin L., Terry N. Overexpression of Glutathione synthetase in indian mustard enhances cadmium accumulation and tolerance, Plant Physiology. 1999; 119: 73–80.

[84] Mckersie B.D., Chen Y., Beus M.D., Bowley S.R., Bowler C., Inze D.,. Halluin K.D, Botterman J. Superoxide dismutase enhances tolerance of freezing stress in transgenic alfalfa (*Medicago sativa* L.). Plant Physiology. 1993; 103: 1155–1163.

[85] Payton P., Webb R., Kornyeyev D., Allen R., Holaday A.S. Protecting cotton photosynthesis during moderate chilling at high light intensity by increasing

chloroplastic antioxidant enzyme activity, Journal of Experimental Botany2001; 52 2345–2354.

[86] Babu V., Bansal K.C. Osmotin overexpression in transgenic potato plants provide protection against osmotic stress, 5th International Symposium on Molecular Biology of Potato, Bogensee, Germany, Aug 2–6. 1998

[87] Maqbool S.B., Zhong H., El-Maghraby Y., Ahmad A., Chai B., Wang W., Sticklen M.B. Competence of oat (*Avena sativa* L.) shoot apical meristems for integrative transformation, inherited expression and osmotic tolerance of transgenic lines containing HVA1, Theoretical and Applied Genetics. 2002; 105 201–208.

[88] Kumar S., Dhingra A., Daniell H. Plastid-expressed betaine aldehyde dehydrogenase gene in carrot cultured cells, roots, and leaves confers enhanced salt tolerance. Plant Physiology. 2004; 135 2843-2854.

[89] Ruidang Q., Mei S., Hui Z., Yanxiu Z., Juren Z. Engineering of enhanced glycine betaine synthesis improves drought tolerance in maize. Plant Biotechnology Journal., 2004; 2 477-486.

[90] Tang W., Peng X., Newton R.J. Enhanced tolerance to salt stress in transgenic loblolly pine simultaneously expressing two genes encoding mannitol-1-phosphate dehydrogenase and glucitol-6-phosphate dehydrogenase. Plant Physiology and Biochemistry., 2005: 43 139-146.

[91] Du N., Liu X., Li Y., Chen S., Zhang J., Ha D., Deng W., Sun C., Zhang Y., Pijut P.M. Genetic transformation of *Populus tomentosa* to improve salt tolerance. Plant Cell Tissue and Organ Culture 2012; 108 181-189.

[92] Molinari H.B.C., Marur C.J., Filho G.C.B., Kobayashi A.K., Pileggi M., Rui P.L. Jnr, Luiz F.P.P., Luiz G.E.V. Osmotic adjustment in transgenic citrus rootstock Carrizo citrange (*Citrus sinensis* Osb. x *Poncirus trifoliata* L. Raf.) overproducing proline. Plant Science, 2004; 167 1375-1381.

[93] Yamada M., Morishita H., Urano K., Shiozaki N., Yamaguchi-Shinozaki K., Shinozaki K., Yoshiba Y. Effects of free proline accumulation in petunias under drought stress. Journal of Experimental Botany, 2005; 56 1975-1981.

[94] Honjoh K., Shimizu H., Nagaishi N., Matsumoto H., Suga K., Miyamoto T., Iio M., Hatano S. Improvement of freezing tolerance in transgenic tobacco leaves by expressing the hiC6 gene. Bioscience, Biotechnology, and Biochemistry. 2001; 65 1796-1804.

[95] Gasic K., Korban S.S. Expression of Arabidopsis phytochelatin synthase in indian mustard (*Brassica juncea*) plants enhances tolerance for Cd and Zn. Planta, 200; 225 1277-1285.

[96] Savitch L.V., Allard G., Seki M., Robert L.S., Tinker N.A, Huner NPA, Shinozaki K, Singh J The effect of overexpression of two brassica CBF/DREB1-like transcription factors on photosynthetic capacity and freezing tolerance in *Brassica napus*. Plant Cell Physiology. 2005; 46 1525-1539.

[97] Furukawa J., Yamaji N., Wang H., Mitani N., Murata Y., Sato K., Katsuhara M., Ma K.T.J. An aluminum-activated citrate transporter in barley. Plant Cell Physiology. 2007; 48 1081-1091.

[98] Ingle R.A., Mugford S.T., Rees J.D., Campbell M.M., Smith J.A.C. Constitutively high expression of the histidine biosynthetic pathway contributes to nickel tolerance in hyperaccumulator plants. Plant Cell, 2005: 17 2089-2106.

Chapter 5

The Science of Plant Tissue Culture as a Catalyst for Agricultural and Industrial Development in an Emerging Economy

S.E. Aladele, A.U. Okere, E. Jamaldinne, P.T. Lyam, O. Fajimi, A. Adegeye and C.M. Zayas

Additional information is available at the end of the chapter

1. Introduction

In the last few decades, the flow of biological discovery has swelled from a trickle into a torrent, driven by a number of new methodologies developed in plant tissue culture, recombinant DNA technology, monoclonal antibodies and micro chemical instrumentation [1] Biological research has been transformed from a collection of single discipline endeavors into an interactive science with bridges between numbers of traditional disciplines. This synergy has made biology the "sunrise field" of the new millennium. The whole gamut of new discoveries in biology and allied sciences can be grouped together under a single umbrella term of "Biotechnology".

Biotechnology has been defined as "any technique that uses living organisms, or substances from these organisms, to make or modify a product, to improve plants or animals, or to develop microorganisms for specific uses" [2]. No society has advanced without deploying appropriate technology in place to set the pace for addressing its major problems. Public investment in relevant technology, the application to industries and capturing of the benefit accrue to it is what sets developed nations apart. Previous reports have shown that there is no National economic growth without proper investment in a right technology which is applied in a Nation. Real solutions to priority on national problems like job creation and poverty alleviation is investment in appropriate technology. This is evident in countries that embraced and adopted biotechnology in past technological revolutions and are practicing on an unprecedented scale. Such countries like India, Cuba and South Africa. The application of biotechnology has greater opportunities for developing countries than previous technologies i.e. greater comparative advantages[3]. Agricultural biotechnology addresses issues such as the production of disease resistant, high yielding and very profitable agricultural ventures in both plants and animals. The world population has grown tremendously over the past two thousand years. In 1999, the world population passed the six billion mark. Latest official current world population estimate, for mid-year 2011, is estimated at 7 billion [4]. The population increase in developing countries constitutes 97% of the global increase [5], and it is estimated that by 2050, 90% of the planet's population will reside in the developing countries of the southern hemisphere. The challenge for the future, therefore, lies in global food security that necessitates a doubling of food production in the next 50 years to meet the needs of the population [6]. Most developing countries yet to fulfill their food production potentials; are especially vulnerable in terms of food security. Plant biotechnology plays a key role in complementing other factors necessary for the improvement of crop production such as the use of agrochemicals, irrigation, plant breeding and farm management to address food security. Plant biotechnology, has three broad fields of study. They include plant tissue culture, genetic engineering and plant molecular markers. These applications range from the simple to the sophisticated and in many cases have been appropriate for use in developing countries [2]. For example, biotechnology techniques such as plant tissue culture have been utilized appropriately for many agronomic and food crops to provide more food and planting materials for farmers.

Micropropagation, popularly known for large-scale clonal propagation, is the first major and widely accepted practical application of plant biotechnology. It is described as the *in vitro* initiation of plant culture, propagation, and rooting under controlled environmental conditions for *ex vitro* establishment in the soil. New contributions to *in vitro* techniques for plant propagation in the last decade have simplified micropropagation technology [7]. This covers a wide range of plants including Agronomic species, economic and forest trees.

In Cassava *(Manihot esculenta) for example,* Tissue culture has made possible the mass production of disease-free and uniform plants. The techniques thus bring farmers the great benefit *of* high- quality planting materials [8].Production of planting materials is indispensable in the overall structure of research for conservation of variety purity and supply of high yielding cassava cultivars to stem multipliers and producers[9]. The possibility of using screen house to maintain *in vitro* cultures and rapidly propagate important vegetative crops with less contamination at a reduced cost was investigated A[10]. *In vitro* propagation has been used to regenerate, establish and conserve both economic trees and forest species through organogenesis and somatic embryogenesis. For instance, *Khayagrandifoliola* is an important species native to West Africa [11]. Khaya wood, African mahoganyas it is commercially called is a high priced wood often used for furniture and construction purposes. With its threatened conservation status [12], micropropagation has been auseful tool for mass propagation of superior stock plants as well as genetic improvement and conservation. For the purpose of conservation and multiplication, a reliable plant regeneration protocol from matured seed embryo of *Khayagrandifoliola was developed* [13]. Another species of economic importance is*Plukenetiaconophora*Mull.Arg. (Family: Euphorbiaceae), formerly known as *Tetracarpidium conophorum* and popularly called African walnut. This tree speciesis a perennial climber of economic importance, an edible species and is used medicinally [14]. A prolific shoot multiplication system (protocol) for *Plukenetia conophora* has been reported [15]. Furthermore, the need to conserve and regenerate recalcitrant species has led to the development of *in vitro* protocols for many recalcitrant species including vegetables. *Telfaria Occidentalis*(fluted pumpkin) is a tropical vegetable grown in West African and widely consumed in tropical regions mainly for its richness in protein [16]. Due to problems associated with the sex of the plant regarding reproduction, the female plant in preferred with respect to leaf and fruit production [17]. *In vitro* culture of *T. occidentalis* under different cytokinins and auxins combination was studied [18]. The commercial use of micropropagation is mainly limited to crops with high value and the commercial utility of this technique for the most important crop species is limited as a result of the large numbers needed annually to start up new farms in addition to high production costs. In order to overcome the problems of conventional micropropagation, protocols have been proposed using bioreactors and liquid medium [19]. Bioreactor system which incorporates a number of features in its design has been used to simplify operation and reduce production costs. Automation, using a bioreactor, is one of the most effective ways to reduce the costs of micropropagation [20].

The Temporary Immersion Bioreactors (TIBs) has been shown to reduce some problems usually encountered in permanent liquid cultures such as hyper-hydricity, poor quality of propagules, andnecessity of transplanting on a solid medium in the elongation and/or rooting stage. In comparison with conventional micropropagation on semisolid medium, TIBs provides a superior mass balance. Indeed in the latter comparison the proliferation rate is higher, labor efficiency is improved and, as a consequence, the cost is reduced [21].

This chapter attempts to give an insight on how methods and applications of *in vitro* technology can serve as a catalyst for both agricultural and industrial development in an emerging economy.

2. Methodology

2.1. *In vitro* propagation of Cassava plantlet *(Manihot esceulenta)*

This study aims at the possibility of using screen house to maintain *in vitro* cultures and rapidly propagate important vegetative crops with less contamination at a reduced cost.

Two genotypes of cassava TMS *188/00106* and TMS *083/00125* were obtained from the Inte~na1ionaf Institute of Tropical Agriculture (IITA) while live different medium were prepared using medium [22] with minor adjustments. as follows:

Treatment 1 (T₁) Liquid only
Treatment 2 (T₂) - liquid with 50% normal agar (2 g/l).
Treatment 3 (T₃) - liquid media with filter paper embedded.
Treatment 4 (T₄) - Media with normal agar (4 g/l).
Treatment 5 (T₅) - liquid media with filter paper projecting out.

The pH was taken and dispensing was done at the rate of 3 ml before autoclaving. The sub culturing was done the following day. One hundred and twenty test tubes were used for each variety with 12 test tubes per treatment. A complete set of 60 test tubes With 5 treatments of 12 replicates was placed in the laboratory while the second set was placed in the screen house at the same day for TMS *188/00106;* the same procedure was adopted the following day for TMS *083/00125.* Data was recorded weekly for 5 weeks before sub culturing. The second generation was observed for only two weeks to ensure the sustainability of the observation made during the first generation. The observation on explants survival was scored on a scale of 0 - 3 as follows:

0 - Dead
1 - Alive but not growing
2 - Growing slowly
3 - Growing very well

There were six parameters recorded during the investigation. These include survival (ate, shoot development, root growth. nodal increase, leave development and increases in height. Survival rate was observed for two weeks only white the other five parameters were scored continuously for the rest three weeks consecutively. Only the screen house explants were subculture after 5 weeks to ensure the sustainability of the findings. The subculture materials from the screen house explants' were also placed in both screen house and culture room (laboratory). The same set of observation was carried out on the responses of the explants to the culture medium and environment as in the first generation explants. The summary in Table 1 indicates that out of the six parameters studied, F-probability on survival is significantly different for all the media used. Observation shows no significant difference on the five treatment for shoot root node, , leaves and height development Although there were some effects on the survival of the explants, the laboratory plantlets grows better in liquid and liquid with filter paper embedded media than when placed in the screen house. This might be due to high temperature recorded at the time of placement (32° - 36°C compared to 22 - 25°C in the laboratory), which indicates an interaction between treatment and environment (Table 1).

SIN	Survival		Shoot		Root		Node		Leaves		Helaht	
	SH	LAB	SH	LAB	SH	LAB	SH	LAB	SH	LAB	SH	LAB
I	1.77	2.41	1.92	1.93	0.73	1.55	2.00	1.93	1.92	1.93	1.92	1.96
11	1.67	1.44	1.56	1.42	1.67	1.33	1.86	1.63	1.67	1.46	1.61	1.33
III	1.02	1.81	1.21	2.04	0.46	1.42	1.46	2.17	1.29	2.00	121	2.04
IV	1.73	1.27	1.83	1.00	1.00	0.75	2.04	1.29	2.00	0.96	1.79	1.00
V	1.73	1.27	1.88	1.13	1.63	1.04	2.13	1.29	1.79	1.13	1.92	1.17
CV	42%		49%		89%		48%		51%		48%	
Lsd	0.38.4		0.4470		0.2987		0.4889		0.470		0.439	
Std	0.193		0.2266		0.5892		0.2479		0.2383		0.2228	

SH - Screen House. LAB - Culture room in the Laboratory.

Table 1. Summary of environmental effect on the *in vitro* growth rate of cassava tissue culture.

Obviously, the laboratory' supports the survival of explants in the liquid medium and liquid medium with embedded filter paper. On the other hand, survival is lowest in the screen house with liquid medium containing embedded filter paper. This suggests that before the explants can be transferred to the screen house, there is need to ensure their survival in the laboratory. For T 2 (liquid with 50% normal agar), T4 (media with normal agar, 4 g/l) and T5 (liquid media with filter paper projecting out), the survival was significantly higher in the screen house than in the laboratory. On the other hand TMS 083/00/25 survived better in the liquid media with embedded filter than TMS 188/00106. The survival rate of TMS 188100106 was also better in liquid medium with 50% normal agar ($2g/l$), media with normal agar and liquid media with filter paper projecting out than TMS 083/00/25. This suggests that for long storage before sub culturing, laboratory may be ideal while for short time storage and immediate rapid mass propagation screen house may be adopted.

Table 2 shows that TMS 188/00106 survived better in liquid media than TMS 083/00125. No significant different between the two genotypes in most of the media except on survival in liquid media only. Figure 1 Shows that screen house plantlet grow relatively uniform for all the five treatments while Figure 2 indicates that plantlets in T1 grew faster than others in the culture room.

It can therefore be concluded that when the need arises, in vitro plantlets of cassava can be raised adequately in the screen house and even be raised faster than the laboratory as long as the temperature does not exceed

SIN	Survival		Shoot		Root		Node		Leaves		Height	
	01	02	01	02	01	02	01	G2	01	02	01	G2
I	2.37	1.51	2.25	1.34	1.29	0.77	2.29	1.39	2.29	1.30	2.21	1.46
11	1.67	1.37	1.46	1.53	1.79	1.05	1.67	1.84	1.54	1.60	1.46	1.47
III	1.13	1.71	1.21	2.04	0.71	1.17	1.25	2.38	1.13	1.17	1.29	1.96
IV	1.54	1.46	1.33	1.50	1.00	0.75	1.79	1.54	1.54	1.42	1.42	1.38
V	1.54	1.46	1.58	1.42	1.00	1.67	1.63	1.79	1.42	1.5	1.54	1.54
CV	43%		50%		92%		48%		52%		51%	
Lsd	0.3898		0.4494		0.5889		0.4791		0.4684		0.4600	
Sed	0.1976		0.2278		0.2985		0.2429		0.2375		0.2332	

G_1=Genotype1=TMS88100106. G_2= Genotype 2 = TMS 083/00125

Table 2. Summary of genotypic effect on the *in vitro* growth rate of cassava tissue culture.

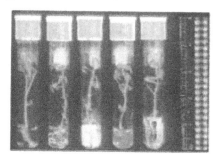

Figure 1. Screen house performance of the 5 treatments of cassava tissue culture. T_1 - Liquid only; T_2• liquid with 50% normal agar (2 g/l); T_3 - liquid media with filter paper embedded; T_4 - media with normal agar (4 g/l); and T_5 - liquid media with filter paper projecting out.

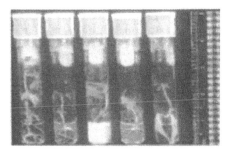

Figure 2. Laboratory performance of the 5 treatments of cassava tissue culture. T₁ - Liquid *only;* T₂ - liquid with 50"10 normal agar (2 gll); T₃ . liquid media with filter paper embedded; T₄ - media with normal agar (4 g/I); and T₅ - liquid media with filter paper projecting out.

2.2. *In vitro* propagation of an endangered medicinal timber species *Khaya grandifoliola* C. Dc.

Considering the fact that this forest tree species seeds are recalcitrant in nature and producing adequate number of seedlings for any meaningful plantation establishment programme from seeds stored for long time is very difficult, this present work aims to describe a reliable plant regeneration protocol from matured seed embryo.

Seeds of *K. grandifoliola* used were collected from the Genebank of National Center for Genetic Resources and Biotechnology (NACGRAB), Ibadan, Nigeria (07°23.048′N 003°50.431′E). Ninety (90) seeds used for this experiment were washed with mild liquid detergent (Tween-20) under running tap water for 10 min. This is followed by surface sterilization in 70% ethanol for 5 min and 0.1% mercuric chloride for 10 min followed by 3 rinses in sterile distilled water. The embryos were carefully excised with ease together with some endosperm attached and then cultured on basal medium supplemented with 3% w/v sucrose, 0.1 g inositol and gelled with 0.7% w/v agar at various concentration of cytokinins and auxin in a 17 ml test tube. The cytokinins used were benzylaminopurine (BAP) and kinetin (KIN), while naphthalene acetic acid (NAA) was the auxin used. All growth regulators were added before autoclaving. The pH was adjusted to 5.7 ± 0.2 and autoclaving was done at 121°C for 15 min. The cultures were incubated in a growth room at 26 ± 2°C under a 16 h photoperiod with cool-white fluorescent light. There were nine treatments, and ten explants were cultured per treatment and later arranged randomly on the shelves in the growth room. After four weeks, the cultures were evaluated for shoot length, root length, number of nodes and number of roots. The data taken were subjected to statistical analysis using SAS/PC version 9.1. The observed means of the characters were subjected to Least Significant Difference (LSD) to show the mean separation.

Data in Table 3 revealed that different concentrations of the cytokinin BAP and the auxin NAA tested in this study had a significant effect on the regeneration of plantlets. The longest shoot length (7.4 mm) was exhibited for explants cultured on MS-medium supplemented with 0.075 mg/L (Kin) + 0.01 mg/L (NAA) and this value is 3 fold higher than that found for embryo cultured on 0.10 mg/L (Kin) + 0.01 mg/L (NAA) whose average shoot length was 2.7 mm. These results showed that the most adequate culture medium for obtaining the longest aver-age root length (7.53cm) per culture after four weeks was MS-medium supplemented with of BAP at 1.0 mg/L plus NAA at 0.1 mg/L, while the shortest root length (1.47 cm) was exhibited by MS-medium supplemented with 0.15 mg/L (BAP) + 0.01 mg/L (NAA), this indicates that increasing the level of auxin (NAA) increases the length of roots and vice-versa. However, the highest number of nodes (4.0) was observed on plantlets cultured on MS-medium supplemented with 1.0 mg/L (KIN) + 0.01 mg/L (NAA).

S/N	Media	Shoot length	Root length	Number of nodes	Number of roots
1	0.125 mg/l (BAP) + 0.01 mg/l (NAA)	4.20	1.53	3.00	1.00
2	0.15 mg/l (BAP) + 0.01 mg/l (NAA)	4.67	1.47	2.00	1.00
3	0.05 mg/l (KIN) + 0.01 mg/l (NAA)	3.7	4.53	2.00	1.00
4	0.075 mg/l (KIN) + 0.01 mg/l (NAA)	7.4	4.20	3.00	1.00
5	0.10 mg/l (KIN) + 0.01 mg/l (NAA)	2.7	4.03	2.00	3.00
6	0.125 mg/l (KIN) + 0.01 mg/l (NAA)	4.92	5.53	3.00	1.00
7	1 mg/l (BAP) + 0.1 mg/l (NAA) + 10 mg/l (adenine sulphate)	5.78	3.50	3.00	3.00
8	1 mg/l (KIN) + 0.01mg/L (NAA) + 10 mg/l (adenine sulphate)	4.82	4.50	4.00	1.00
9	1 mg/l (BAP) + 0.1 mg/l (NAA)	5.87	7.53	3.00	2.00
	LSD	0.14	0.13	0.00	0.00

Table 3. Effect of plant growth regulators on shoot length, root length, number of nodes, and number of roots regeneration from embryo culture of *K. grandifoliola*.

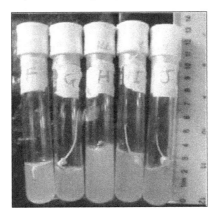

Figure 3. The growth stages of *K. grandifoliola* through embryo culture *in vitro* (two weeks after culture).

These findings are in agreement with those reported earlier [23] on Cacti (*Pelecyphora aselliformis*) and *Nealolydia lophophoroides*; the work on *Aloe barbebsi* [24] and *Turbinicapus laui* [25] indicate that using a high concentration of BAP and NAA in different concentrations was a limiting factor for shoot formation and increases root formation. The result of this study showed that the optimum medium for regeneration of *K. grandifoliola* MS-medium supplemented with 1.0 mg/L (BAP) + 0.1 mg/L (NAA) + 10 mg/L adenine sulphate because the values obtained for all the parameter measure was moderately high and optimum. The fact that the number of roots increased to 3 in medium 5 and 7 could be due to the increase in the concentration of NAA to 0.1 mg/L. It has been established that auxins like NAA increases the root formation in the presence of low cytokinins [26].

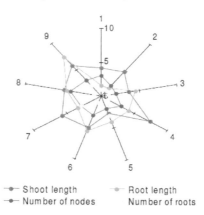

Figure 4. The effect of different levels of kinetin and BAP on the performance of *K. grandifoliola*.

2.3. *In vitro* culture of *Telfairia occidentalis* under different cytokinins and auxin combinations

The aim of this work was to investigate the *in vitro* regeneration potential of *Telfairia occidentalis* under different hormonal combination. The nodal cuttings collected from four weeks old seedlings raised in the screen house at NACGRAB were used as explant. The explants were surface-sterilized in 15% NaOCl + 2 drops of Tween 20 per 100 ml for 25 min. They were then cut with a sharp sterile knife into single node cuttings. Three or four explants from a seedling were cultured on the prepared medium to which either naphthalene acetic acid (NAA) or benzylaminopurine (BAP), indolebutyric acid (IBA), indole-3-acetic acid (IAA) and kinetin had been added. Different concentrations were investigated for each of the auxins and cytokinins. The basal medium used comprised of Murashige and Skoog macro and micro-elements, vitamins 3% sucrose, 10 mg/L ascorbic acid, 0.1 g/L myo-inositol, and 0.02 g/L cysteine. Cultures were incubated in the dark at 25 ± 2oC for duration of six weeks for shooting and rooting induction. The number of roots and nodes were counted and recorded on the sixth week.

Among all the growth hormones used, IBA (0.05 mg/L) + BAP (0.01 mg/L) combination gave the best result for both rooting and shooting while the highest number of nodes was observed in BAP (0.05 mg/L) + NAA (0.01 mg/L). The application of kinetin both in combination with NAA and alone resulted in premature senescence with lower number of nodes. This is in agreement to the findings of [27] who showed that kinetin is not a suitable hormone for regeneration of *Telfairia* especially if it will be kept *in vitro* for a long time. However BAP (0.05 mg/L) + IAA (0.01 mg/L) combination resulted in lowest number of nodes and MS alone produced callus without regenerating into a plantlet. The result shows that *in vitro* growth of *T. occidentalis* is hormone specific.

2.4. In vitro micro-propagation of *Plukenetia conophora* Mull.Arg

This study describes a reliable and prolific shoot multiplication system (protocol) for *Plukenetia conophora.* Excised embryos and nodal cuttings from growing seedlings served as the major explants used for the study. Matured fruits were collected from the field gene bank of NACGRAB, Ibadan.

2.4.1. Dis-infection of explants

Nodal cuttings: The nodal segments were obtained from the stems of actively growing seedlings, washed with liquid detergent under running tap water and disinfected using standard disinfection procedures before culturing. For *embryos,* cotyledons obtained from the nuts were reduced into small size, washed with liquid detergent under running tap water and then disinfected appropriately.

2.4.2. Culture conditions

The culture media consisted of MS basal medium supplemented with vitamins, myo-inositol, sucrose, casein hydroxylate and growth regulators. The pH of the medium was adjusted before autoclaving. All the cultures were kept at 24 ± 2 ^{0}C under cool light fluorescent lamp for a photoperiod of 16 hours.

2.4.3. Shoot induction from matured embryos

Matured excised embryos were cultured on shoot induction medium supplemented with different concentrations of KIN (0.0 – 0.50) mg/l with NAA /0.05 mg and BAP (0.0 – 0.50) mg/l with NAA 0.05mg,whileexcised nodal segment from actively growing stem were cultured for direct organogenesis on MS medium supplemented with 0.0-0.45mg/l BAP/0.05mg NAA and 0.0-0.45mg/l KIN/0.05mg NAA for shoot proliferation.

The studies showed that the excised embryos regenerated *in vitro* after12 weeks of culture had a healthy appearance. *In vitro* regeneration was achieved on MS basal medium without growth regulators. On MS basal media fortified with growth regulators, the best mean result of shoot length was recorded on medium supplemented with 0.3mg KIN and 0.01mg NAA. The medium augemented with 0.3mg KIN and 0.05mg NAA gives the longest root length. These concentrations induced a higher percentage of explants with shoots and shoot number per explant than the hormone-free treatment. Therefore, the introduction of growth regulators led to the increase in shoot length and number of nodes. However, an increase in the concentrations of NAA from 0.01mg to 0.05 mg resulted in a decline in the number of shoots formed and an increase in the root length (Table 6).

In the *in vitro* regeneration of nodal cuttings, nodal culture on MS basal (the hormone-free treatment) showed no significant growth until supplemented with cytokinin, hence the need for supplementing the MS basal medium with cytokinin. MS medium containing 0.30mg BAP and 0.05mg NAA gave the best mean shoot length, number of shoots, and node number. An increase in the concentration of BAP above 0.30mg/l led to a decline in shoot length. The addition of Casein hydroxylate improved the shoot response as observed in the difference in growth response of nodal cuttings on MS containing 0.2mg/l BAP and casein hydroxylate and 0.2mg/l BAP without casein hydroxylate as there were differences in their shoot response.

Media	Shoot length (cm)	Root length (cm)	No. of nodes
MS only	4.97±0.03	3.90±0.05	2.00±0.00
MS + 0.30mg KIN + 0.01mg NAA	3.10±0.00	2.00±0.24	3.00±0.00
MS + 0.30mg KIN + 0.05mg NAA	2.00±0.00	5.50±0.24	2.00±0.00
MS + 0.40mg BAP + 0.01mg NAA	2.00±0.00	5.50±0.24	2.00±0.00

Mean result ±standard error.

Table 4. Effect of KIN, BAP and NAA on *in vitro* regeneration of *Plukenetia conophora* embryos after 12 weeks.

Media	Shoot length (cm)	No. of shoots	No. of nodes
MS only	No significant growth		
MS + 0.20mg BAP (no casein hydroxylate)	0.57±0.03	1.00±0.00	1.00±0.00
MS + 0.20mg BAP + 0.05mg NAA	0.70±0.08	1.00±0.00	1.00±0.00
MS + 0.30mg BAP + 0.05mg NAA	2.13±0.05	2.00±0.00	2.00±0.00
MS + 0.35mg BAP + 0.05mg NAA	1.10±0.05	2.00±0.00	2.00±0.00
MS + 0.40mg BAP + 0.05mg NAA	0.83±0.07	1.50±0.29	1.67±0.27
MS + 0.45mg BAP + 0.05mg NAA	0.63±0.13	1.30±0.27	1.25±0.25
MS + 0.30mg KIN + 0.05mg NAA	0.53±0.02	1.00 ±0.00	1.00 ±0.00

Table 5. Effect of BAP, KIN and NAA on *in vitro* regeneration of *Plukenetia conophora* nodal explants.

3. The Potential of Temporary Immersion Bioreactors (TIBs) in scaling up crop production, to meet agricultural demand in developing countries

Temporary Immersion Bioreactor system (TIBs) is a relatively recent micropropagation procedure that employs the use of automated gadgets to control rapid multiplication of plant cultures under adequate conditions. TIBs provide a more precise control of the adequate conditions (gaseous exchange, illumination etc.) required by plants for growth, development and survival than the conventional culture vessels. This bioreactor system incorporates a number of features specifically designed to simplify its operation and reduce production costs.

TIBs consist of three main phases: Multiplication, Elongation and rooting phase. Plantlets propagated in TIBs have better performance than those propagated by conventional methods of micropropagation. This is as a result of a better handling of the *in vitro* atmosphere and the nutrition. The system also provides a rapid and efficient plant propagation system for many agricultural and forestry species, utilizing liquid media to avoid intensive manual handling. In addition to diminishing production costs regarding labour force, Temporary Immersion Bioreactors save energy, augment micropropagation productivity and efficiency.

3.1. Use of bioreactor technology?

Bioreactors provide a rapid and efficient plant propagation system for many agricultural and forestry species, utilizing liquid media to avoid intensive manual handling. Several authors have reported the use of bioreactors for plants propagation [28],[29]. To reduce the intensive labour requirement along with the production cost during plant propagation by tissue culture technique, there is an immense need of developing scale-up systems and automation [30]. This method for large scale production of plants is promising at industrial level. Employing bioreactors with liquid medium for micropropagation is advantageous due to the ease of scaling-up [31]. Large-scale plant propagation using bioreactor can also be beneficial in terms of year round production of the propagules of useful plants resulting in comparatively less labour cost and time [32]. The major advantages of using bioreactor culture system for micropropagation of economically important plants includes the potential for scaling-up in lesser time limit; Reduction in the production cost as well as an automated control of physical and chemical environments during growth phase of the plant cultures.

3.2. Importance of TIBs technology to agricultural development in emerging countries

Modern biotechnology has put the micropropagation industry on the verge of exciting new breakthroughs. It offers improvements in virtually every area of crop production and utilization, with potential benefits to agriculture, the food industry, consumers and the environment. As the world's population continues to grow, it is anticipated that there could be many mouths to feed in the next few decades. The advances made possible through micropropagation (TIBs) will be essential to meet global food needs by increasing the yield, quality and quantity of crops available to farmers. TIBs offer further benefits in form of non-food crops. Through mass propagation of specific economic species, it will be possible to arrest desertification, soil erosion in affected areas and also increase industrial crop production as renewable sources of medicines, industrial chemicals, fuels etc. They offer potential benefits to the commercial farmers, industries, public, research scientists and students. The potential benefits of TIBs are summarized below.

- Mass propagation of agronomic food crops to enhance food security. (i.e All year round production and supply of planting materials to farmers).
- Scaling up of the production of specific crops for industrial use (A step towards commercialization) e.g pineapple juice.

- Mass propagation of economic tree species (e.g *Eucalyptus spp., Adzadiracta indica, Accacia sp*) for addressing environmental problems like desertification and erosion.
- Job creation.
- Inspire collaborations among institutions on specific economic and ecological projects.

3.3. The use of TIBs in the mass propagation of Plant Genetic Resources (Using Pineapple, plantain, sugarcane and *Eucalyptus sp* as a case study)

The objectives of this work were:

- To produce high planting density of crops through an efficient and rapid production system to meet conservation and large scale farming production demands.
- To produce homogenous plantlets for research and development purposes.

Four major stages are recommended for effective mass propagation of plant cultures using temporary immersion systems, these include:

3.3.1. Stage 1: Collection and Establishment of the mother explants on agar gel medium.

The establishment of truly aseptic cultures usually involves the following sequential steps:

Step 1.   Pre-propagation step or selection and pre-treatment of suitable plants.

The mother plants are selected and screened before transporting to the green house environment. The health status of the donor mother plant and of the plants multiplied from it are among the most critical factors, which determine the success of a tissue culture operation. Hence, indexing of the mother plants for freedom from viral, bacterial, and fungal diseases is a normal procedure before undertaking propagation in large-scale plant propagation through tissue culture [33]. This step is crucial as it tends to reduce the microbial load present at the time of collection and which may hinder or interfere with the *in vitro* processes.

Step 2.   Initiation of explants - surface sterilization, establishment of mother explants.

This involves the sequential disinfection of the mother plant under aseptic conditions, culture initiation and establishment on a suitable growth media. The process requires excision of tiny plant pieces and their surface sterilization with chemicals such as ethyl alcohol, sodium hypochlorite and repeated washing with sterile distilled water before and after treatment with chemicals. The appropriate growth media for each crop was prepared. The pH was adjusted to 5.7 ± 0.2 before autoclaving at $121°C$ for 15 min and culture initiation was carried out under the laminar flow hood.

The initiated cultures were then transferred to the growth room and incubated at $26 \pm 2°C$ under a 16 h photoperiod with cool-white fluorescent light.

Step 3.   Subculture of explants on agar gelled media for multiplication and proliferation.

This involves the subculture of established explants on agar gelled media with a specific auxin/cytokinins combination to induce proliferation. In this step, explants were cultured on the appropriate media for multiplication of shoots. The primary goal was to achieve propagation without losing the genetic stability. Repeated culture of axillary and adventitious shoots, cutting with nodes, somatic embryos and other organs from Stage I led to multiplication of propagules in large numbers. The propagules produced at this stage were further used for multiplication by their repeated culture.

3.3.2. Stage 2: Zero shelving of plants to liquid medium.

Sometimes it is necessary to subculture the *in vitro* derived shoots onto different media for elongation and rooting for *ex vitro* transfer. However, if cultures must be mass propagated in Temporary Immersion System, they must be allowed to pass through the zero phase upon establishment and when they have gained a proliferation capacity/potential on the agar gelled multiplication media. This is usually done in order to prepare the explants for proper adaptation, survival and desired *in vitro* response in the next phase which utilizes only liquid medium.

3.3.3. Stage 3: Introduction and cultivation of ex-palnts into Temporary Immersion Bioreactors.

This Phase refers to Plant cell/tissue growth and development in liquid medium under the control of Temporary immersion systems. It utilizes the advantages of liquid medium coupled with automated control of culture conditions to rapidly multiply explants thereby increasing exponentially the multiplication coefficient of the explants. Only healthy *in vitro* derived shoots that successfully passed through the zero shelving were introduced in TIBs set up. For the set up at NACGRAB, a total of 6 temporal immersions by pneumatic driven medium transfer were made daily. The immersion frequency was 3 minutes at 3 hours interval with a pre-immersion and post immersion period of 10 minutes respectively. This stage involves 3 major phases **(Figure 5a, b & c).**

In vitro response of plantlets in each of the 3 phases of TIBs is highly dependent on certain factors including hormonal combination, duration of each feeding cycle and the overall timing/duration to which it is subjected to. eg. higher cytokinins (BAP) concentration to auxins in the multiplication media, Gibberillic acid (GA_3) for elongation and auxins (IAA, NAA and IBA) for rooting.

i. *Multiplication phase-* Plantlets were cultured on MS liquid medium void of agar with the appropriate di-hormonal combination depending on the plant species. Pineapple and plantain were transferred to MS liquid medium + 4.0mg/l –BAP and 1.8mg/l –NAA, while Eucalyptus and sugarcane were transferred to MS liquid medium + 0.5mg/l-BAP and 0.1mg/l –NAA, for a period of 8 -12 weeks respectively.

ii. *Elongation phase-* This phase aims at developing proliferating buds into plantlets that are lengthy, strong and robust enough to stand acclimatization and withstand adverse conditions during *ex vitro* transfer. To achieve this, Plantlets were cultured in MS liquid medium void of agar with 1.0g/l of Gibberellic acid. After 21 days, plantlets were removed and placed in a rooting medium.

iii. *Rooting and harvest-* In order to induce rooting, elongated plantlets were cultured in a liquid MS media containing Auxin treatments of (A) 0.5mg/l- IBA, (B) 1.0mg/l-IBA (C) 0.5mg/l- NAA (D)1.0mg/l –NAA (E) A combination of 0.5 mg/l NAA and 0.5 mg/l IBA and (F) A combination of 1.0mg/l- IBA+ and 1.0mg/l-NAA for 4 weeks respectively. All media had equal volume in the same culture vessel. At the end of 4 weeks, plantlets were harvested by an initial disinfection of the mouth of culture bottle with 1% sodium hypochlorite. Bottle was opened, plantlets carefully collected.

(a)   (b)

(c)

Source: Lyam *et al.*, 2012.

Figure 5. (a) Multiplication (proliferation in clusters), (b) Elongation (Stem growth elongation), (c) Rooting (well developed root system) of pineapple

The timing to achieve the desired goal at each phase varies depending on the individual species ability to respond to each phase accordingly. The common feature to all the phases is the use of liquid media (void of agar) to aid nutrient uptake and automation. Transfer of plantlets from one step to the next is carried out aseptically under the laminar flow hood.

Clumps of shoots derived were separated after rooting and not before as this usually cause
tissue wounding and stimulate the exudation of phenolic compound which interferes with
the physicochemical factors that trigger root formation. In this way, multiplied plantlets
were elongated and rooted to produce complete plants and harvested. Harvest is carried out
by an initial disinfection of the mouth of culture bottle with 1% Sodium hypochlorite. Bottle
was opened and plantlets carefully collected.

3.3.4. Stage 4: Acclimatization and Ex vitro transfer

This is the final stage of the tissue culture operation including the use of bioreactor after
which the micro propagated plantlets are ready for transfer to the greenhouse. Steps are
taken to grow individual plantlets capable of carrying out photosynthesis. Collected
plantlets were sorted and prepared for acclimatization based on their sizes and rooting
capacity. *In vitro* micro propagated plants are weaned and hardened. The hardening of the
bioreactor propagated plantlets is done gradually from high to low humidity and from low
light intensity to high intensity conditions. Rooted plants were washed with tap water and
acclimatized *ex vitro* on a medium composed of Coconut fibre, Top Soil and Stone dust
mixed in the ratio 7:2:1 which can be left in shade for 3 to 6 days where diffused natural
light conditions them to the new environment. The plants were transferred to an
appropriate substrate for gradual hardening Figure 15 -20.

Figure 6. Potted and acclimatized plants in the screen house at NACGRAB.

These stages are universally applicable in large-scale multiplication of plants. The individual
plant species, varieties and clones require specific modification of the growth media,
weaning and hardening conditions. A rule of the thumb is to propagate plants under
conditions as natural or similar to those in which the plants will be ultimately grown *ex-
vitro*. Micropropagated plants must be subjected to an adequate duration of time required
for their proper hardening.

Figure 7. TIBs set up at the NACGRAB, Ibadan, Nigeria

3.4. Challenges of temporary immersion bioreactor systems

The use of liquid cultures in bioreactor for plant propagation imposes several problems such as leakage of endogenous growth factors, the need for an initial high concentration of the inoculum, lack of protocols and production procedures, increased hyperhydricity and malformation, foam development, shearing and oxidative stress, release of growth inhibiting compounds by the cultures and contamination. Unfortunately culture contamination which is a major problem in conventional commercial micropropagation is even more acute in bioreactors [34]. In conventional micropropagation, discarding a small number of the contaminated vessels is an acceptable loss; in bioreactors, even a single contaminated unit is a huge loss. However, despite these difficulties, a number of commercial laboratories have developed effective procedures to control contamination in bioreactors. Highlighted below are some of the challenges [35], [36], [37].

3.4.1. Inadequate protocols and production procedures

Protocols for proliferation on semi solid media are not always efficient when used in bioreactors. However, as no one protocol is utilized for all species, it becomes quite difficult to achieve success at a goal. Development of protocol for scaling up cultures in bioreactors entails extensive research and development in all phases of TIBs (multiplication, elongation and rooting). It is possible to record success at one phase and not overcome the challenges at the next phase. For the efficient scaling up of cultures in temporary immersion bioreactors for commercialization, protocol for multiplication, elongation and rooting must be developed.

3.4.2. Increased hyperhydricity and malformation

The major disadvantage encountered when plants are cultured in liquid media is the problem of shoot malformation. Plants tend to accumulate excess of water in their tissue resulting to anomalous morphogenesis, a phenomenon known as Hyperhydricity. The plants that develop in liquid media are fragile, have a glassy appearance, with succulent leaves or shoots and a poor root system [38]. Hyperhydricity in micropropagation has been reported in previous studies [39], [40].

3.4.3. Problems of foaming, shear and oxidative stress

Growth and proliferation of the biomass in bioreactors depends on airflow supply for the aeration and mixing, and for the prevention of the plant biomass sedimentation. In many plants cultivated in bioreactors, continuous aeration, mixing, and circulation cause shearing damage, cell wall breakdown, and accumulation of cell debris, which is made up mainly of polysaccharides.

The problem of foaming and shear damage of tissues including their potential solutions in bioreactors has been reported [41], [40].

3.4.4. Release of growth inhibiting compounds by the cultures

This is also known as the *in vitro* Phenolic browning or oxidation. The presence of phenolic compounds which cause death of explants has been another important problem of micropropagation especially in woody perennials, in addition to various bacterial and fungal infection. Some of these exudates appear as a reaction to injury and/or infection. In tissue culture they appear after tissue excision and are many times aggravated by growth media constituents [42]. The release of growth inhibiting compounds by *in vitro* cultures has been reported [43], [44]

Some of the solutions to this problem as suggested are as follows:

- Addition of activated charcoal (0.2-3.0% w/v) to the medium ,
- Addition of polymeric polyvinylpyrrolidone (PVP) or polyvinylpolypyrrolidone (PVPP) to the medium. These absorb phenols through hydrogen bonding.
- Additions of anti-oxidants or reducing agents like citric and ascorbic acids, thiourea glutathione and L-cysteine in the medium or before surface sterilization. These reduce the redox potential of explants and stop the oxidation reactions (Marks and Simpson, 1990) [45].
- Addition of diethyl-dithiocarbonate (DIECA) (2g.l-1) in the rinses after surface sterilization and as droplets at the time of micro grafting.
- Addition of amino acids like glutamine, arginine and asparagine to the media.
- Reduction of salt concentration in the growth media. Others may include:
- Frequent subcultures onto fresh media.
- Use of liquid medium for easier and quicker dilution of toxic products.

- Reduction of wounded tissues to decrease exudation.
- Soaking of explants in water before culturing to reduce browning.
- Incubation of fresh cultures in darkness for the first few days of culture.

The suggestions above have provided solution to phenolic oxidation in micropropagation and are widely employed in most laboratories across the globe.

3.4.5. Microbial contamination

After three decades of research and development in plant tissue culture, microbial contamination by yeasts, fungi, bacteria, viruses, mites and thrips are still the major problem that has hampered the establishment of truly aseptic plants and their successful Micro - propagation in bioreactors. The influence of bacteria on shoot growth can range from total inhibition to no apparent effect. The contaminating bacteria and fungi may be endophytic or epiphytic, pathogenic or saprophytic [46]. Another type of hazard for plant tissue and cell cultures is caused by 'latent' bacteria and viruses that do not produce any symptoms on the plant or any visible growth on the medium for long periods of time *in vitro* even after several subculture cycles; microbial contamination in culture has been reported [47].

3.4.6. Control of contamination

Prevention of contamination in bioreactors requires a proper handling of the plant material, equipment and cultures during transfers and production. Only the surface sterilized explants, multiplied in small vessels and indexed for freedom from diseases are used to initiate cultures in bioreactors. If the bioreactor is small, it is sterilized in an autoclavable plastic bag, sealed with a cotton wool plug, and opened only under the laminar flow cabinet. Despite the precautions taken in initiating cultures, bioreactors can become contaminated from the environment or from latent microbes in the culture. The contamination can be controlled with one or a combination of anti-microbial compounds, acidification of the media, and micro-filtration of the medium [48]. While most of the fungal and bacterial diseases are eliminated during surface sterilization and culture, viruses and viroids survive through successive multiplication if the mother plant is infected [49].

Author details

S.E. Aladele, A.U. Okere, E. Jamaldinne, P.T. Lyam and O. Fajimi
Biotechnology Unit, National Centre for Genetic Resources and Biotechnology (NACGRAB), Moor Plantation, Ibadan, Oyo State, Nigeria

A. Adegeye
Department of Forest Resources Management, University of Ibadan, Ibadan, Oyo State, Nigeria

C.M. Zayas
Genetic and Biotechnology Department, National Institute for Sugarcane Research, Boyeros, Havana City, Cuba

4. References

[1] Crouch, J. H., Vuylsteke, D. and Ortiz, R. (1998). Perspectives on the application of biotechnology to assist the genetic enhancement of plantain and banana (*Musa* spp). *Electronic Journal of Biotechnology* (1):6-10.

[2] Kelly, V., Adesina, A.A., Gordon, A., 2003. Expanding access to agricultural inputs in Africa: a review of recent market development experience. *Food Policy* (3): 379–404.

[3] Bodulovic, G. (2005). Is the European attitude to GM products suffocating African development? *Functional Plant Biology.*(32): 1069–1075.

[4] Rosenberg (2012) http://geography.about.com/bio/Matt-Rosenberg-268.htm

[5] Swaminathan, M.S. (1995). Population, environment and food security. *Issues in Agriculture*, No 7. CGIAR, Washington DC.

[6] James C. (1997). Progressing public-private sector partnership in International Agriculture Research and Development. In: *ISAAA* Briefs No 4, p. 1-32.

[7] Lyam P.T.,Musa M. L., Jamaleddine, Z.O., Okere, A.U., Carlos A. and Odofin, W.T. (2012). The potentials of Temporary immersion bioreactor TIBS in meeting crop production demand in Nigeria. *Journal of biology and life science* ISSN 2157 -6076 Vol. 3:1.

[8] Dalmacio I. F. (1992). Biotechnology for sustainable Agriculture National Institute of Biotechnology and Applied Microbiology, University of the Philippines.Dalziel, J. M. 1937. *The useful plants of West Tropical Africa*. Whitefriars Press, London, 612 pp.

[9] Acheapong E. (1982). Multiplication and Conservation of Cocoyam (*Xarthosoma sajiltifoljum*) germplasm by application of tissue culture methods Ph.D Thesis. University of Burmingham, UK.

[10] Aladele S. E. and Kuta D. (2008). Environmental and genotypic effects on the rowth rate of *in vitro* cassava plantlet (*Manihot esceulenta*) African Journal of Biotechnology Vol. 7 (4), pp. 381-385.

[11] Gbile Z.O.(1998). Collection, Conservation and Utilization of medicinal plants. In Robert PA, & Janice EA (eds.). Conservation of Plant Genes III: Conservation and Utilization of African Plants. Missouri Botanical Gardens Press. pp. 163-174.

[12] IUCN (2006). IUCN Red List of Threatened Species. IUCN, Gland, Switzerland.

[13] Okere, A. U. and Adegeye, A. (2011). *In vitro* propagation of an endangered medicinal timber species *Khaya grandifoliola* C. Dc. *African Journal of Biotechnology* Vol. 10(17), pp. 3335-3339,

[14] Dalziel J.M. (1937) *The useful plants of West Tropical Africa. Whitefriars press,London. 612pp.*

[15] Fajimi O. and Fashola T. R. (2010). *In vitro* micro-propagation of *Plukenetia conophora* Mull.Arg. Journal of South Pacific Agriculture. 14(1&2):1-5

[16] Fagbemi T.N.F., Eleyinmi A.F., Atum H. N., Akpambang O. (2005) Nutritional composition of fermented fluted pumpkin (Telfairiaoccidentalis) seeds for production of "ogiriugu". IFT Annual Meeting, July 15-20 – New Orleans, Louisana.

[17] Akoroda M. O. (1990). Ethno botany of *TelfariaOccidtendalic*Among Igbos of Nigeria. Econ. Bot. 44 (1): 29-39.

[18] Sanusi I. S., Odofin W. T,, Aladele S. E, , Olayode M. N., Gamra E. O., Fajimi O. (2008) In vitro culture of Telferia occidentalis under different cytokinins and auxin combination. African journal of biotechnology. 7 (14):2407-2408

[19] Teisson, C. and Alvard, D. (1995). *A new concept of plant in vitro cultivation in liquid medium: Temporary Immersion. In: Current Issues in Plant Molecular and Cellular Biology.*Terzi, M., Cella, R. and Falavigna, A. (Editors). Kluwer Academic Publisher pp 105-109.

[20] Levin, R., Stav, R., Alper, Y. and Watad, A. A. (1998). A technique for repeated non-axenic subculture of plant tissues in a bioreactor on liquid medium containing sucrose. *Plant Tissue Culture Biotechnology.* (3):41–45.

[21] Escalona, M., Lorenzo, J.C., Gonzales, B.L., Daquinta, M., Borroto, C.G., Gozales, J.I. and Desjardine, Y. (1999). Pineapple (*Ananascomosus* L. Merr.) micropropagation in temporary immersion system. *Plant Cell Report.* (18): 743–748.

[22] Murashige T, Skoog F. (1962). A revised medium for rapid growth and bioassays with tobacco tissue culture. Physiol. Plant, 15: 473-497.

[23] Bustamante M. A., Heras M. G. (1990). Tissue culture of Cacti species. XXIIIrd International Horticultural Congress, Firenze (Italy) Aug-27-sept-1, 1990 contributed paper (oral) No. 1344 p. 163.

[24] Feng-F, Li-Hong B, Lu-Qing F, Xie-Jian Y, Fen F, Li HB, Lu QF, Xie J. Y. (2000). Tissue culture of *Aloe* spp. J. Southwest Agric. Univ. 22(2): 157-159.

[25] Mata-Rosas M, Monroy-de-la-Rosa MA,Goldammer KM, Chavez-Avilla VM, Monroy-de-la-Rosa M. A. (2001). Micropropagation of *Tubinicapuslaui*glass et Foster, an endemic and endangered species. *In vitro* Cell. Dev. Biol. Plant, 37(3): 400-404.

[26] Youssef E. M. S. (1994). Effect of cytokinins and reported cub cultures on *in vitro* micropropagation potentiality of *Acacia salicina*Lindll. Tissue culture Lab.; TimberTrees Res. Dept., Inst., Agric. Res. Cen., Giza, (1): 30-43.

[27] Balogun M. O., Ajibade, S. R, Ogunbodede B. A. (2002). Micropropagation of fluted pumpkin by enhanced axillary shoot formation. Niger. J. Hort. Sci. 6: 85-88.

[28] Escalona, M., Samson, G., Borroto, C. and Desjardins, Y. (2003). Physiology of Effects of Temporary Immersion Bioreactors on Micropropagated Pineapple Plantlets. *In Vitro Cell Developmental Biology of Plant.* (39): 651-656.

[29] Etienne, H., Dechamp, E., Etienne, B. D. and Bertrand, B. (2006). Bioreactors in coffee micropropagation. *Brazilian Journal of Plant Physiology.* (18): 2-3.

[30] Aitken-Christie, J. (1991). Automation In: Micropropagation, Debergh, P.C., Zimmerman, R. H. (Editors). Kluwer Academic Publishers. Dordrecht, Netherland. pp. 342-354.

[31] Preil, W. (1991). *Application of bioreactors in plant propagation. In: Micropropagation: Technology and application.*Debergh, P.C. and Zimmerman, R.H (Editors.), Kluwer Academic Publishers, Dordrecht,The Netherlands. Pp. 425–455.

[32] Preil, W., Florek, P., Wix, U. and Beck, A. (1988). Towards mass propagation by use of bioreactors. *ActaHorticulturae.* (226):99–106.

[33] Schmidt, J. Wilhem E. and Savangikar,V.A. (2002). Disease detection and elimination. In: Low cost options for tissue culture technology in developing countries. *Proceedings of a Technical Meeting, Joint FAO/IAEA Division of Nuclear Techniques in Food and Agriculture.* Pg 55-61.

[34] Leifert, C., and Waites, W. M. (1998). Bacterial growth in plant tissue culture. *Journal of Applied Bacteriology*, (72):460–466.

[35] Leifert, C. and Waites, W.M. (1990). *Contaminants of plan tissue culture.* Internl. Assoc. Plant Tiss. Cult. Newsl. (60): 2-13.

[36] Leifert, C. and Woodward, S. (1998). Laboratory contamination management: the equirement for microbiological quality assurance. *Plant Cell Tiss. Org. Cult,* (52): 83-88.

[37] Leifert, C. (2000). Quality Assurance Systems for Plant Cell and Tissue Culture: The problem of Latent Persistance of Bacterial Pathogens and *Agrobacterium* Transformed vector systems. *In. Proceedings of the International Symposium, methods and markers for quality assurance in micropropagation.* A.C. Cassells et al. (Eds.) Acta. Horticulturae. (560):Pp. 530.

[38] Etienne, H., Lartaud, M., Michaux-Ferrie`re, N., Carron, M.P., Berthouly, M. and Teisson, C. (1997). Improvement of somatic embryogenesis in *Heveabrasiliensis (*Mu¨ ll. Arg.) using the temporary immersion technique. *In Vitro Cellular and Developmental Biology of Plant.* (33): 81–87.

[39] Etienne, H., Dechamp, E., Etienne, B. D. and Bertrand, B. (2006). Bioreactors in coffee micropropagation. *Brazilian Journal of Plant Physiology.* (18): 2-3.

[40] Ziv, M., Ronen, G. and Raviv, M. (1998). Proliferation of meristematic clusters in disposable presterlized plastic bioreactors for large scale micropropagation of plants. *In vitro Cell Developmental Biology of Plant.* (34): 152-158.

[41] Scragg, A. H. (1992). Large-scale plant cell culture: methods, applications and products. *Current Opinion Biotechnology.* 3:105–109.

[42] Seneviratne, P. and Wijesekara, G.A.S (1996). The problem of phenolic exudates in in vitro cultures of mature Heveabrasiliensis. *Journal of Plantation Crops,* 24(1):54-62.

[43] Fowler, M. R. (2000). Plant cell culture, laboratory techniques. In: *Encyclopedia of cell technology,* Spier, R.E (Editor), Wiley, New York. Pp. 994–1004.

[44] Pierik, R.L.M. (1987). In *Vitro Culture of Higher Plants.* Kluwer Academic Pulishers, Dordrecht, pp. 25-33. http://dx.doi.org/10.1007/978-94-009-3621-8.

[45] Marks, T.R. and Simpson, S.E. (1990). Reduced phenolic oxidation at culture initiation *in vitro* following the exposure of field grown stock plants to darkness or low level irradiation. *Journal of horticultural science.* (65):103-111.

[46] Deberge, P.C and Maene, L.J. (1981). A scheme for commercial propagation of ornamental plants by tissue culture. *ScientiaHorticulturae.,* 14: 335-345.

[47] Debergh, P. C. and Read, P.E. (1991). Micropropagation. In: *Micropropagation Technology and Application.* Debergh, P.C, and Zimmerman, R.H. (Editors). Kluwer Academic Pulishers,Dordrecht, pp. 1-13. http://dx.doi.org/10.1016/0304-4238(81)90047-9

[48] Schmidt, J. Wilhem E. and Savangikar,V.A.(2002).Disease detection and elimination. In: Low cost options for tissue culture technology in developing countries. *Proceedings of a Technical Meeting, Joint FAO/IAEA Division of Nuclear Techniques in Food and Agriculture.* Pg 55-61.

[49] Sessitsch, A., Reiter, B., Pfeifer, U. and Wilhelm, E. (2002). Cultivation-independent population analysis of bacterial endophytes in three potato varieties based on eubacterial and Actinomyces-specific PCR of 16rRNA genes. *FEMS Microb. Ecol.,* 39: 23-32. http://dx.doi.org/10.1111/j.1574-6941.2002.tb00903.x

Chapter 6

Somaclonal Variation in Tissue Culture: A Case Study with Olive

A.R. Leva, R. Petruccelli and L.M.R. Rinaldi

Additional information is available at the end of the chapter

1. Introduction

Micropropagation of woody plants and fruit crops constitutes a major success in the commercial application of in vitro cultures. An important aspect to be considered when deriving perennial plants from micropropagation is the maintenance of genetic integrity with regard to the mother plant. In this regard, somaclonal variation has been reported at different levels (morphological, cytological, cytochemical, biochemical, and molecular) in micropropagated plants [1]. The economic consequence of somaclonal variation among regenerated plants is enormous in fruit crops and woody plants, because they have long life cycles. In consequence, the behaviour of micropropagated plants should be assessed after their long juvenile stage in field conditions. The occurrence of somaclonal variation is a matter of great concern for any micropropagation system. In order to evaluate its presence several strategies were used to detect somaclonal variants, based on one or more determinants from among morphological traits, cytogenetic analysis (numerical and structural variation in the chromosomes), and molecular and biochemical markers [2]. In addition, studies on somaclonal variation are important for its control and possible suppression with the aim of producing genetically identical plants, and for its use as a tool to produce genetic variability, which will enable breeders the genetic improvement. Somaclonal variation has been studied extensively in herbaceous plants, whereas few studies have focused on temperate perennial fruit crops.

This chapter provides a survey of the technical approaches for identifying somaclonal variation in perennial fruit crops and is intended to provide a synthesis of the literature on the topic. In addition, recent advances in the characterization and detection of somaclonal variation in olive plants produced *in vitro* by nodal explants and somatic embryogenesis are reported in detail.

2. Somaclonal variation

In nature, the genetic diversity and variability within a population are generated via recombination events. Factors such as natural selection, mutation, migration and population size influence genetic variability in different ways. In 1958 a novel, artificially produced, source of genetic variability was reported [3], the higher plant cells cultured *in vitro* showed a genetic instability that was also characteristic of regenerated cells. The first observation of somaclonal variation was reported [4]. Subsequently, the variability existing in plant tissue and cell cultures received much attention and neologisms were proposed by Larkin and Scowcroft [5] to refer to the results of *in vitro* cultures of plants. The term 'somaclone' was coined to refer to plants derived from any form of cell culture, and the term 'somaclonal variation' was coined to refer to the genetic variation among such plants. The growth of plant cells *in vitro* and their regeneration into whole plants is an asexual process that involves only mitotic division of the cells. In this context, the occurrence of uncontrolled and random spontaneous variation when culturing plant tissue is a major problem [6]. *In vitro*, the conditions of culture can be mutagenic and regenerated plants derived from organ cultures, calli, protoplasts and somatic embryos sometimes can show phenotypic and genotypic variation [7]. Some, or all, of the somaclones may be physically different from the stock donor plants [8]. Usually, variability occurs spontaneously and can be a result of temporary changes or permanent genetic changes in cells or tissue during *in vitro* culture. Temporary changes result from epigenetic or physiological effects and are nonheritable and reversible [9]. In contrast, permanent changes are heritable and often represent expression of pre-existing variation in the source plant or are a result of *de novo* variation [5]. The literature to date indicates that somaclonal variation can range in scope from specific trait to the whole plant genome. Somaclonal variation provides a valuable source of genetic variation for the improvement of crops through the selection of novel variants, which may show resistance to disease, improved quality, or higher yield [10, 11, 12, 13].

3. Origin and causes

Although somaclonal variation has been studied extensively, the mechanisms by which it occurs remain largely either unknown or at the level of theoretical speculation in perennial fruit crops [14,15]. A variety of factors may contribute to the phenomenon. The system by which the regeneration is induced, type of tissue, explant source, media components and the duration of the culture cycle are some of the factors that are involved in inducing variation during *in vitro* culture [16].

3.1. Regeneration systems

Regeneration systems can be ranked in order from high to low in terms of genetic stability, as follows: micropropagation by preformed structures, such as shoot tips or nodal explants; adventitiously derived shoots; somatic embryogenesis; and organogenesis from callus, cell and protoplast cultures [17, 18]. Cellular organization is a critical factor for plant growth, whereas *in vitro* the loss of cellular control, which gives rise to disorganised growth, is a characteristic of somaclonal variation [6,19].

Although the direct formation of plant structures from meristem cultures, without any intermediate callus phase, minimises the possibility of instability, the stabilising influence of the meristem is sometimes lost *in vitro* cultures [6].

Somatic embryogenesis and enhanced axillary branching are the methods used most extensively in commercial micropropagation systems [20]. Somatic embryogenesis has the potential to produce the greatest number of plantlets in a short time, and makes possible the use of bioreactors for the large-scale production of somatic embryos [21] and their delivery through encapsulation into artificial seeds [22, 23, 24, 25]. Enhanced axillary branching involves the abolition of apical dominance to achieve the de-repression and multiplication of shoots, and has become a very important method on account of the simplicity of the approach and rapid propagation rate [26, 27]. These methods are considered to produce genetically uniform and true-to-type plants, because the organised meristems generally are believed to be immune to genetic changes [28, 29]. Several reports of experimental studies support this view [30, 31, 32, 33, 34, 35]. However, there is an increasing body of evidence that indicates that in embryogenic cultures, selection in favour of 'normal cells' does not always take place during development and that growth of mutant cells can occur as well, which can induce variability in the cultures [36].

3.2. Explant source

Genetic fidelity largely depends on explant source [37]. The explant tissue can affect the frequency and nature of somaclonal variation [38,39]. The use of meristematic tissues, such as the pericycle, procambium and cambium, as starting materials for tissue culture reduces the possibility of variation [40]. In contrast, highly differentiated tissues, such as roots, leaves, and stems, generally produce more variants, probably due to the callus-phase, than explants that have pre-existing meristems [41]. Furthermore, preparation of many explants from only one donor plant increases the possibility of variation in cultures [42]. This illustrates the importance of the donor plant with respect to its inherent genetic composition and genome uniformity in any of its components. Somaclonal variation can arise from somatic mutations already present in the tissues of the donor plant [6]. To test for pre-existing somaclonal variation, the somatic embryos obtained in the first round of regeneration may be subjected to another round of *in vitro* regeneration. Tissues that show pre-existing variation should yield more variability in the first somaclonal generation than in the second generation, and thereafter the variation in the second round can be eliminated or stabilised [15].

3.3. Medium components

The hormonal components of the culture medium are powerful agents of variation. The effect of the type and concentration of plant growth regulators on the incidence of somaclonal variation in different plant species remains a topic of debate. Unbalanced concentrations of auxins and cytokinins may induce polyploidy, whereas under a low concentration or total absence of growth regulators the cells show normal ploidy [43]. In addition, rapid disorganised growth can induce somaclonal variation [6]. Sub- and supra-optimal levels of growth regulators, especially synthetic compounds have been linked with somaclonal variation [44, 45]. Auxins added to cultures of unorganised calli or cell suspensions increase genetic variation by increasing the DNA methylation rate [46]. Similarly, in callus cultures of strawberries, the presence of the synthetic auxin 2,4-Dichlorophenoxyacetic acid (2,4-D) is often associated with genetic abnormalities, such as polyploidy and stimulation of DNA synthesis, which may result in endoreduplication [47, 48, 49, 50]. It would seem that growth regulators preferentially increase the rate of division of genetically abnormal cells [51]. High levels of cytokinins do not directly affect the rate of somaclonal variation in the banana cultivars 'Nanjanagudu Rasabale' and 'Cavendish'; in this contest it would seem that the genotype has the greatest effect on somaclonal variation [52, 53]. Conversely, high levels of benzyladenine (BA) cause the number of chromosomes in the banana cultivar 'Williams' to increase [54]. In addition, diphenylurea derivatives are implicated in the incidence of somaclonal variation in bananas [55].

3.4. Duration and number of culture cycles

The frequency of somaclonal variation increases as the number of subcultures and their duration increases, especially in cell suspensions and callus cultures [56, 57, 58]. Moreover, the rapid multiplication of a tissue or long-term cultures may affect genetic stability and thus lead to somaclonal variation [59, 56, 60]. A statistical model has been proposed for predicting the theoretical mutation rate, primarily on the basis of the number of multiplication cycles [61]. However, the model has limited application, due to the complexity of biological systems.

3.5. Effect of genotype

Conditions of culture *in vitro* can be extremely stressful for plant cells and may initiate highly mutagenic processes [62, 63]. However, different genomes respond differently to the stress-induced variation, which indicates that somaclonal variation also has genotypic components. In callus cultures of strawberry, the genotype and type of explant strongly influenced the occurrence of somaclonal variation [64]. The differences in genetic stability are related to differences in genetic make-up, because some components of the plant genome may become unstable during the culture process, for example the repetitive DNA sequences, which can differ in quality and quantity between plant species [65]. In banana tissue culture the most important factor that influenced dwarf off-type production was found to be the inherent instability of the cultivars; for example, the cultivar 'New Guinea Cavendish' showed a higher level of instability *in vitro* than 'Williams'. The dwarf off-types remained stable during *in vitro* culture, and the conditions under which tissue was cultured that induced dwarfism did not induce reversion of the dwarf off-type trait [66]. In *Musa* species, the type and rate of variation was specific and depended on genotype [67, 68] and genome composition [40]. An interaction between genotype and the tissue culture environment is also reported [69].

4. Genetic changes that contribute to somaclonal variation

During plant growth and development *in vivo*, gross changes in the genome can occur during somatic differentiation, including endopolyploidy, polyteny and amplification or diminution of DNA sequences [70]. The processes of dedifferentiation and redifferentiation of cells may involve both qualitative and quantitative changes in the genome, and different DNA sequences may be amplified or deleted during the cell reprogramming. In addition, this process is related closely to the tissue source and the regeneration system [65]. Gross and cryptic chromosomal changes, or extensive changes in chromosome number, occur early during induction in an *in vitro* culture [5]. Variation in chromosome numbers and structures, and chromosome irregularities (such as breaks, acentric and centric fragments, ring chromosomes, deletions and inversions) are observed during *in vitro* differentiation and among regenerated somaclones [71, 72, 5]. Such rearrangements in chromosomes may result in the loss of genes or their function, the activation of genes previously silent, and the expression of recessive genes, when they become haploid. The irregularities in the chromosomes may be lost during plant regeneration and result in the production of 'normal' plants, or appear in the regenerated somaclones.

Cryptic changes, such as point mutations, are also expected to occur and may affect the chloroplast or mitochondrial genomes. In addition, transpositional events, such as the activation of transposable elements, putative silencing of genes and a high frequency of methylation pattern variation among single-copy sequences, play a role in somaclonal variation [73, 74].

The tissue culture environment may result in the modification of DNA methylation patterns [62, 75]. Global methylation levels and methylation of specific sites are documented in several crops, e.g. oil palm [76], grapevine [77, 78] and apple [79]. In addition, epigenetic changes, such as DNA methylation and histone modifications, may be associated with the physiological responses of the plant cells to the conditions *in vitro* [75]. Several epigenetic systems have been studied: variation for morphological traits, such as flower colour and shape, leaf colour and shape, and plant height; resistance to disease; and maturity date [80]. The rate of these changes varies not only in response to tissue culture conditions, but also among species and even among cultivars of the same species [81].

There are several extensive reports of morphological and genetic variation of several plant species, primarily herbaceous species, but few studies have been conducted on perennial fruit crops, which indicates that knowledge of somaclonal variation in these plants is lacking [1]. Currently, many markers are available to verify the fidelity of perennial fruit crops at the morphological, physiological and molecular levels.

In light of the many factors that can lead to somaclonal variation, the characterisation of micropropagated plants is essential to help to modify the protocol/s with which genetically true-to-type plants are obtained, so that the procedures can be used with predictable results.

5. Morphological, cytological and biochemical markers

Morphological markers usually are used to identify species, genera and families in germplasm collections. Somaclonal variants can be detected easily by morphological characteristics, such as plant height, leaf morphology and abnormal pigmentation [68]. For example, a sweet cherry (*Prunus avium*) somaclonal variant was characterised by morphological parameters, namely evaluation of plant vigour, leaf morphology, stomatal density, photosynthesis activity, the formation of floral buds, and the size, shape and colour of the fruit [82].

Chromosomal alteration and ploidy changes are highlighted by cytogenetic analysis, including chromosome counting and/or flow cytometry. Cytometry has been used to identify the particular characteristics of somaclonal variation in *Vitis vinifera* [83, 84, 85] and *Citrus lemon* [7]. Proteins and isozymes have been used widely as markers for identifying cultivars and characterising somaclonal variation in many fruit species [33]. Isozyme analysis has been used to assess genetic fidelity in *Citrus* plants regenerated through organogenesis, and somatic embryogenesis [86].

6. Molecular markers

In some instances, discrepancies between molecular markers and phenotypic data are observed [87]. These discrepancies relate to the complexity of the plant genome, and the markers that are normally used often cannot give a complete view [88]. To resolve this difficulty, assay at the molecular level should be examined in relation to morphological changes. Currently, different molecular analytic techniques have used to point out somaclonal variation in tissue culture and in regenerants of several plants. Randomly amplified polymorphic DNA (RAPD), amplified fragment length polymorphism (AFLP), simple sequence repeat (SSR) and inter-simple sequence repeat (ISSR) markers have been used to study the genetic fidelity or genetic variability in micropropagated fruit crops.

RAPD markers are suitable for detecting somaclonal variation in *Prunus persica* and the variation observed is genotype-dependent [89]. In contrast, RAPD fingerprinting confirmed genetic fidelity in microcuttings of *Cedrus libani* [90], micropropagated plants of *Cedrus atlantica* and *C. libani* [91], plants regenerated from leaf explants of *Citrus sinensis* cv. Bingtangcheng and cv. Valencia [92], and in *Citrus limon* plants [7]. Moreover, RAPD analysis showed clonal stability in micropropagated plants of *Pyrus* [93], and in plants of hazelnut (*Corylus*) regenerated from long-term *in vitro* cultures [94]. Polymorphic RAPD marker have been observed in apple and pear cultivars regenerated from adventitious shoots [95, 96], whereas no polymorphism was observed in apple and pear cultivars regenerated from vegetative shoot apices [96]. No polymorphism in RAPD markers was observed between plants propagated *in vitro* and donor plants of clones of the hybrid *Castanea sativa* × *C. crenata* [97].

AFLP markers have been used to assess both genetic variation among axillary shoots and shoots regenerated from leaf- and petiole-derived calli of kiwifruit (*Actinidia deliciosa*). [98].

A close relationship between the type of explants, regeneration system, and the presence of somaclonal variation was detected in shoots regenerated from leaf callus [99].

Plants of *Vitis vinifera* cvs. Mission and Valerien regenerated from callus cultures showed polymorphism using methylation-sensitive restriction enzymes [100].Variation in DNA methylation has been highlighted in apple plants micropropagated from *in vitro* axillary shoot cultures [101].

SSR markers were applied to different *V. vinifera* cultivars regenerated from shoot cultures and from anther and ovary embryogenic callus lines. Homogeneous amplification profiles were revealed in the *in vitro* samples and donor plants [102]. In addition, SSR markers revealed genetic stability in *V. vinifera* cv. Crimson Seedless shoot cultures [103] and in micropropagated plants of the apple rootstock Merton 793 [104].

In general, the use of one type molecular marker to assess the stability of *in vitro* propagated plants may be insufficient. Recently, several authors used multiple molecular marker types to study somaclonal variation in regenerants of several plant species. In *Actinidia deliciosa* cultures, a relatively low level of polymorphism was detected with RAPD markers, whereas with SSR markers the level of polymorphism detected was higher [105]. Genetic stability was analyzed in plantlets of almond (*Prunus dulcis*) regenerated by axillary branching with RAPD markers and confirmed by ISSR analysis [106]. The genetic fidelity of plantlets obtained by indirect somatic embryogenesis from anthers and ovaries of *V. vinifera* cv. Grignolino and cv. Dolcetto was detected by SSR and AFLP analysis [107].

7. Olives

Olives belong to the genus *Olea* (family Oleaceae). The genus comprises 35 species, of which the most important is *O. europaea*, which comprises two subspecies subsp. *sylvestris* (wild olive) and subsp. *europaea* (cultivated olive). Olive cultivars are subdivided into two types: i) olive-oil cultivars, which are used for the production of olive oil, in which the ripe fruit contains at least 20% oil; and ii) table cultivars, which produce fruits with a lower oil content and are destined for canning. Olive cultivation is concentrated in southern Europe (Spain, Italy, France, Portugal and Greece) and the Mediterranean region (Turkey, Syria, Libya, Morocco and Tunisia). The olive is the longest-cultivated crop in the Mediterranean basin and the most economically important fruit tree in this part of the world. Over the last few decades, cultivation has expanded to other countries, such as South Africa, Argentina, Chile, the USA (in California), Australia and China. Currently, 9.4 million ha of olive orchards worldwide produce about 16 million tons of olive fruits that are processed into 2.5 million tons of oil and 1.5 million tons of table olives. For centuries, olive plants have been propagated vegetatively by suckers, grafting and cuttings. *In vitro* culture offers a novel alternative method of propagation to these traditional approaches.

We report the studies conducted on olive plants produced *in vitro* by nodal explants and somatic embryogenesis to verify their fidelity to donor plant.

7.1. Regeneration systems for olive

The ability of a single cell to divide and give rise to a whole plant (cellular totipotency), is the theoretical and experimental basis of modern plant biotechnology. Perennial and woody plants generally are considered to be recalcitrant in culture and are difficult to regenerate [108]. In commercial regeneration systems, the two methods that are used most commonly for olive plants are micropropagation and somatic embryogenesis.

7.1.1. Micropropagation

Schaeffer [109] defined micropropagation as the *in vitro* clonal propagation of plants from shoot tips or nodal explants; in the case of olive, nodal explants are used. Indirect systems, such as the differentiation of adventitious shoots after a phase of disorganised callus formation, are used less frequently for perennial plants because of the possible selection of several cell-lines in callus. The multicellular origin of adventitious buds is considered to generate a high risk that the regenerated plants will lose fidelity to the parent plant. In a commercial setting, this threat is often serious enough to eliminate any further consideration of micropropagation as a cloning method.

In vitro vegetative propagation by nodal explants allows rapid production of a large number of plants. The procedure consists of the following stages: culture establishment, shoot proliferation, rooting and acclimatization *extra vitro* (Figure 1). Therefore, the degree of success depends on the genotype, selection of explants, media and culture conditions, and levels of growth regulator. The micropropagation of olives was first achieved a quarter of a century ago [110, 111, 112]. However, progress in improving the technique has been relatively slow, due to the inherently slow growth of olive explants. The rate of proliferation of olive explants is limited by the low frequency of bud sprouting and poor growth rates of secondary shoots [113, 114, 115]. In recent years, several studies have investigated and improved the technique to define an *in vitro* propagation sufficiently effective for large-scale commercial application [116].

Figure 1. Stages of micropropagation by nodal explants of olive plants following from left to right: Stage 1= shoots proliferation; Stage 2= rooting of shoots; Stage 3 = acclimatization of shoots *ex vitro*

7.1.2. Somatic embryogenesis

This regeneration system has been well documented in several species, using a wide range of plant tissues as explant sources. Immature embryos are suitable for induction of embryogenesis. This capability is, in most instances, not merely an intrinsic property of a species and instead is under genetic control, such that individual genotypes within a species can differ in their ability to undergo somatic embryogenesis. The first report of somatic embryogenesis in olive used a portion of cotyledons from immature embryos [117, 118] (Figure 2). Subsequent studies have focused on the induction of somatic embryogenesis in different olive cultivars [119, 120, 121]. Improved protocols have enabled the induction of somatic embryogenesis from mature tissues (petioles) obtained from shoots grown *in vitro* [119] portions of the radicle and cotyledons [122], ovaries, stamens, leaves and petioles [123,124], and cell suspension cultures derived from mature olive tissue [125].

 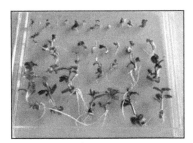

Figure 2. Somatic embryos of cv Frangivento at different development stages

7.1.3. Propagation by nodal explants and somatic embryogenesis compared

Micropropagation has a significant advantage over somatic embryogenesis in that it is thought to reduce the potential for undesirable somaclonal variants among the regenerated plants, whereas in somatic embryogenesis the risk of genetic instability is high. Somatic embryogenesis has an advantage over micropropagation in that it generates a new plant with both root and shoot meristems from actively dividing somatic cells in the same step and within a short time period, whereas micropropagation requires additional steps and a longer time frame.

7.2. Field performance of olive plants raised *in vitro*

Despite the commercial importance of clonal fidelity of plants produced by tissue culture, the field performance and genetic integrity of olive plants regenerated from *in vitro* cultures is reported sporadically [126, 127, 128]. These factors are of paramount importance for olive cultivars. Olive trees have a long life span (hundreds of years), a long juvenile period, broad genetic diversity and consequent variability of the fruit, which may influence traits that pertain to the composition and quality of the oil, such as aroma and taste [129]; hence, any variation of the genotype may modify the olive oil characteristics. From this point of view it is critical that the morphological and molecular performance of mature plants, derived from axillary buds and somatic embryogenesis, should be evaluated.

7.2.1. Morphological characterisation of plants derived from nodal explants

Morphological and biological characteristics have been used widely for descriptive purposes and are commonly used to distinguish olive cultivars [130]. A morphological approach is the main initial step when describing and classifying olive germplasm (131). So, a morphological analysis was used initially to evaluate the plants produced by micropropagation (termed 'micropropagated plants'; MPs) using the protocol reported in [113,132] and in [116].

The donor plant (DP) for the initial explants for micropropagation was a 20-year-old tree of *Olea europaea* cv. Maurino. The same donor plant, multiplied by cuttings, was used as the control plant in the field evaluation. In 1998, 70 16-month-old MPs cultured in pots and 20 control plants (Cp) were transferred to an experimental field situated in San Pancrazio, Firenze, Italy (43°39'36.00" N, 11°11'25.80" E). Twenty-four morphological parameters for vegetative and reproductive traits were considered (Table 1).

During field growth, no significant differences in most of the vegetative parameters were noted between the MP and Cp plants. The data revealed no differences in growth habit, vegetative growth, and canopy and trunk area. Only the leaves and drupes of MPs were slightly wider than the Cp plants but still retained the characteristic leaf and drupe shapes of the cultivar. The productivity of fruiting shoots was similar, despite the slightly different number of flowers per inflorescence; in olive fruit set only occurs in 2–5% of the flowers, so the slight difference did not have any effect on fruiting [133]. Pit traits were identical between the MP and Cp plants (Figure 3). Olive oil was extracted from both plant groups and subjected to sensory tests and chemical composition analysis. The oils had a strong fragrance and there was no variability in composition between the oil produced by the MP and Cp plants (Figure 4). Thus, the MP plants showed morphological and productive uniformity with the Cp plants in terms of vegetative, reproductive and oil traits. Any genetic variation present among the MPs was unrelated to development, vegetative growth and production quality. These results are related to the protocol used for *in vitro* culture. It is important to stress the absence of relevant somaclonal variation because, as reported by Rani and Raina [1], a procedure for micropropagation should be released for commercial application only when analyses on mature plants have established that the procedure does not induce undesirable somaclonal variation.

Vegetative characters	Inflorescence and fruit characters
HP Plant height: measured in meter from the soil level to the highest point	*IL Inflorescence length in mm
CP Canopy projection to the soil: measured at the two widest diameters in m2	NF Number of flowers per inflorescence

Vegetative characters	Inflorescence and fruit characters
VP Canopy volume in m^3	NO Number of olive fruits per fruiting shoot
TA Trunk area in cm^2	FL Fruit length in mm
VSG Vegetative shoot growth in cm	FW Fruit width in mm
VSN Node number of vegetative shoots	FL/W Fruit length/width
VSI Internode length of vegetative shoots in cm	FFW Fruit fresh weight in g
*FS Number of feather shoots (lateral shoots developing from axillary buds formed in same year) on the vegetative shoots	FDW Fruit dry weight in g
*FG Feather shoot growth in cm	PL Pit length in mm
*FN Feather shoot node number	PW Pit width in mm
*FI Internode length of feather shoots in cm	PL/W Pit length/width
LBL Leaf blade length in mm	PFW Pit weight in g
LBW Leaf blade width in mm	FFW/PW Fruit weight/Pit weight
BL/W Blade length/width	FY Production weight in Kg
LA Leaf area in mm^2	
*LFW Leaf Fresh weight in mg	
*LDW Leaf Dry weight in mg	
*DW Dry weight mg per 100 mm^2	

*no used for microplants

Table 1. Quantitative descriptors of olive plants propagated through nodal explants, and somatic embryogenesis observed in field cultivation

Figure 3. Micropropagated olive orchard

Figure 4. Olive oil extracted by micropropagated plants

7.2.2. Molecular characterisation and assessment of genetic fidelity of plants derived from micropropagation

At the end of the second year of field cultivation, an initial molecular analysis was conducted on five MP, Cp plants and the donor plant. Twenty-one 10-base primers were used for PCR-RAPD analysis. The primers generated a total of 182 amplification fragments. All of the primers produced monomorphic amplification patterns in the MPs, and no differences were found in the amplification pattern among MPs, Cps and donor plant.

In 2006, 12 randomly chosen 9-year-old mature MPs were evaluated by means of RAPD and ISSR analysis. Total genomic DNA was extracted from fresh leaves of the MPs and donor plant using the DNeasy Plant Mini Kit (Qiagen, Hilden, Germany). To detect any genetic changes, the RAPD and ISSR results were compared among all MPs, and between MP and Cp plants. The 40 RAPD primers generated 301 scorable band classes, whereas the 10 ISSR primers produced 46 reproducible fragments. The RAPD primer amplification products were monomorphic in MPs and donor plant, as well as the ISSR primers produced monomorphic bands within MPs, and between MP samples and donor plant [134] (Figure 5).

The molecular analyses, undertaken in two separate studies, at different plant ages, and using two types of molecular markers, supported the genetic stability and uniformity of *Olea europaea* cv. Maurino MPs. In addition, these results confirmed the reliability of the morphological analysis results [127, 128].

7.2.3. Morphological characterisation of plants regenerated by somatic embryogenesis

Plantlets were obtained from embryogenic tissue induced from immature cotyledons of the cultivar Frangivento using the methods reported in reference [118], (Figure 2). Fifty somatic seedlings were recovered from embryogenic long-term cultures (3 years), planted in small commercial pots (1.5 l) and placed in a greenhouse in 1992. The survival frequency was 83% after 3 months. The somatic seedlings and MPs from Frangivento donor plant, which was used as a putative control (Pc), were grown in large pots in a greenhouse up to 1997. During cultivation in pots, the somatic plants showed developmental behaviour related to growth rate and habit that differed from that of the MPs. In 1998, 43 somatic plants and 10 Pc plants were transferred to field condition in San Donato, Firenze, Italy (43°33'46.08" N, 11°10'21.00" E). In the field, the somatic plants retained the different developmental behaviour observed in pots. The plants were monitored for several years, during which time it was possible to detect morphological variation related to potential yield, inflorescences, fruits and their characteristics. In total, 32 morphological traits were analysed (Table 1). On the basis of the preliminary observations of plant growth in pots, in the field we recorded, in particular, data for two variant phenotype groups classified as bush olive somaclones (BOS; four representative trees) and columnar olive somaclones (COS; four representative trees) (Figure 6).

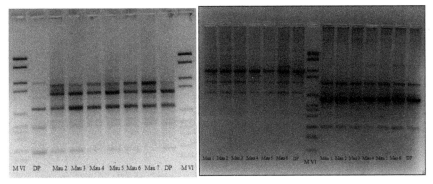

Figure 5. RAPD-PCR amplification patterns for micropropagated plants cv Maurino (Mau) and donor plant (DP) using primers left:M10 right SAT3 and 40148. MVI markers

Figure 6. Somaclonal olive plants: BOS and COS phenotypes, left: somaclones during the growth in pots in greenhouse, right: mature somaclones grown in experimental field

Four replicates for each phenotype were chosen in agreement with the number of replicate plants in an olive germplasm collection, according to the International Union for the Protection of New Varieties of Plants descriptor list. The relationships among the BOS and COS groups and Pc plants were investigated by analysis of variance for morphological traits and by a multivariate method (cluster analysis).

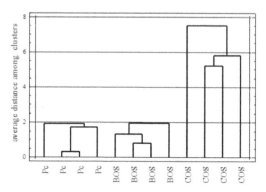

Figure 7. Dendrogram based on morphological traits of the BOS, COS groups and Putative control (Pc) according to a hierarchical clustering.

Wide phenotypic variation in shoot growth responses was observed between the BOS and COS groups. The differences were related to the growth rate, leaf, fruit and pit traits. The architecture of the COS plants, which is determined by plant height, canopy projection and canopy volume, differed from that of Pc plants. In the BOS plants, the reduction in plant height, increase in feather shoot number, and reduction in the dimensions of the leaves, inflorescences and fruits, jointly contributed to the more compact growth habit than that of the Pc plants. Cluster analysis was able to separate and characterise the BOS and COS groups from the Pc plants, as expected, on the basis of the different growth habits and dimensions of the leaf, fruit, pit and inflorescence (Figure 7) [128].

Figure 8. Somaclonal olive plants, left: flowering, right: fruiting

The morphological variations detected among the somaclonal plants were positive and important variations for a possible future utilization of somaclones as rootstocks or new genotypes, and were not accompanied by deleterious changes in other agronomically or horticulturally important traits (Figure 8).

Our results on somaclonal variation among olive plants produced by somatic embryogenesis are in agreement with preliminary data recorded for field-grown juvenile olive plants derived from embryogenic callus of the cultivar Moraiolo [135].

7.2.4. Molecular characterisation of somaclonal olive plants

Bush olive somaclone (BOS), columnar olive somaclone (COS) and Frangivento donor plants grown in the field were analysed with RAPD molecular markers. Total genomic DNA was extracted from fresh leaves using the DNeasy Plant Mini Kit (Qiagen, Hilden, Germany). As a first step, 40 primers (decamers) were used [136], of which 20 primers showed reproducible and well-resolved bands. All subsequent analyses were conducted using these 20 primers.

The 20 RAPD primers generated 198 scorable band classes that ranged in size from 2200 to 210 bp; the number of primer bands varied from seven (CD11 and OPA01) to 13 (OPP10, AH30 and OPP12). Both somaclone types and Frangivento shared a large proportion (86%) of RAPD markers, which suggested the occurrence of homology among the somaclones and Frangivento. These results were expected because all regenerants were derived from the same seed (two cotyledon portions). In addition, the results suggest that genetic changes occurred during the long-term maintenance of embryogenic cultures *in vitro*.

Some primers showed polymorphism between the somaclones and the donor plant, whereas others were polymorphic between the BOS and COS groups. Eight primers (OPP15, OPP10, AH29, AG1, OPP12, OPA07, OPP14 and OPP02), showed a close relationship with growth habit (Table 2). The type of polymorphism detected was either presence or absence of a fragment. The primer OPP15 amplified a single, intense band of approximately 537 pb in COS plants, but this band was absent in BOS plants and Frangivento. In addition, the primer OPP10 amplified two specific bands: a single band of approximately 603 pb present only in BOS plants, and a specific band of approximately 425 pb present in COS plants. OPP14 amplified a specific band present in all somaclones but absent in Frangivento donor plant. Eight polymorphic primers amplified seven specific bands (OPP10$_{603}$, AH29$_{1190}$, AG1$_{550}$, OPP12$_{1350}$, OPA07$_{407}$, OPA07$_{340}$ and OPP02$_{750}$) for BOS plants and three specific bands (OPP15$_{537}$, OPP10$_{425}$ and OPP02$_{800}$) for COS plants.

Primer	Fragments (pb)	COS	BOS	DP
OPP15	1000	–	+	+
	537	+	–	–
	375	–	+	+

Primer	Fragments (pb)	COS	BOS	DP
OPP10	1698	–	–	+
	603	–	+	–
	425	+	–	–
AH29	1190	–	+	–
	1135	+	–	+
	800	+	–	+
AG1	550	–	+	–
	260	+	–	+
OPP12	1170	–	+	+
	1350	–	+	–
	832	–	–	+
	692	–	–	+
OPA07	1114	+	–	+
	1072	+	–	+
	517	+	–	+
	407	–	+	–
	340	–	+	–
OPP14	1230	–	+	+
	725	+	+	–
	625	+	–	+
	210	+	–	+
OPP02	1260	+	–	+
	800	+	–	–
	750	–	+	–
	530	+	–	+

Table 2. Polymorphic fragments scored among the Columnar genotype (COS), Bush genotype (BOS), and Donor Plant (DP)

In 2004, a second RAPD analysis, conducted using the same 20 primers considered in the initial molecular analysis, was performed on 12-year-old somaclonal plants (the plant age refers to the time from transfer to *ex vitro* conditions). The RAPD analysis of the mature olive plants (Figures 9,10) confirmed all patterns of differences between the somaclones and donor plant obtained in the initial study. Moreover, the bands specific to BOS and COS were confirmed. Therefore, when reproducibility is strictly controlled, the RAPD system still seems to be the most rapid and inexpensive for testing variability among plants obtained by somatic embryogenesis. This study indicated clearly that the somaclonal variation in olive was stable and confirmed the morphological results obtained in previous years .

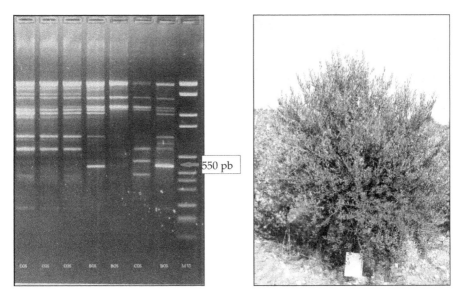

Figure 9. RAPD-PCR amplification patterns for somaclonal olive plants BOS using AG1 primer. Polymorphic DNA fragment is identified by an arrow

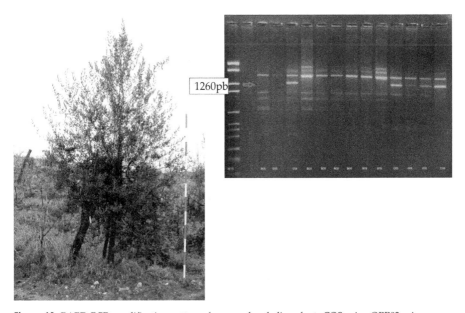

Figure 10. RAPD-PCR amplification patterns for somaclonal olive plants COS using OPP02 primer. Polymorphic DNA fragment is identified by an arrow

7.2.5. Description of morphological variation among the vegetative progeny of somaclonal plants

In view of the stringent requirements for the adoption of novel olive genetic material by working breeders, the variation detected must be hereditable, either sexually or asexually, through conventional vegetative propagation. The aim of our studies on somaclones has been to determine whether the variation in habit is also expressed in the progeny of somaclonal olive plants (daughter plants) obtained by vegetative propagation (rooted cuttings). Data for both the rooting capacity of somaclonal plants and its behaviour under field conditions are reported for 2005 and 2009 respectively.

Type of habitus	Rooting (%)	Root number	Root length (cm)
Columnar (COS)	64.41 n.s	5.5 n.s	2.45 n.s
Dwarf (BOS)	61.15 n.s	4.8 n.s	2.57 n.s

Means were discriminated using Tukey's multiple range test at the 5% level of significant

Table 3. Rooting percentages of the BOS and COS somaclonal olive plants

The BOS and COS variant phenotypes showed a medium-high rooting capacity, with no statistical differences for the parameters studied between the two types of growth habit (Table 3).

Furthermore, in 2006, under an agreement with the Centro di Ricerca e Sperimentazione in Agricoltura (CRSA) "Basile Caramia" Locorotondo, Taranto, Italy, the propagated daughter somaclonal plants were planted in an experimental field at Vivai Conca d'Oro at Palagianello, (40°34'43" N, 16°58'31" E, m. 32 a.s.l.) to verify both the heritability of the habit variation and the behaviour of habit variation in very different environmental conditions. This approach was based on the premiss that the performance of BOS and COS plants might depend not only on their inherent growth habit, but also on the environmental conditions in which they were grown. The environment could limit or enhance the expression of morphological growth phenotypes.

Type of habitus	Plant Height m	Trunk diameter cm	Length of lateral shoots cm	Node number of lateral shoots	axillary shoot number	Node number of axillary shoots
Columnar (COS)	3.2a	11.1ns	76.1a	39.5a	1.9c	5.9b
Dwarf (BOS)	2.4b	10.0ns	65.9b	31.8b	11.8a	11.6a
Putative control	3.0a	9.4ns	55.8c	27.6b	6.0b	5.7b

Means were discriminated using Tukey's multiple range test at the 5% level of significant

Table 4. Comparison of daughter somaclonal olive plants of columnar and dwarf: plant height, trunk diameter, length and node number of lateral and axillary shoots

Preliminary assessment of the performance of somaclonal daughter plants, conducted for 3 years, indicated that morphological characteristics were comparable to those of the somaclonal mother plants. As in the mother plants, the main morphological patterns were the plant height for the BOS phenotype, and length, node number and number of axillary shoots for both phenotypes, in comparison to the putative control (Table 4). In addition, in relation to the environmental conditions at the planting site, the growth performance of the somaclonal daughter plants depended only on the inherent growth habit (Figure 11).

Figure 11. Daughter somaclonal olive plants in experimental field at CRSA

8. Conclusion

Although most perennial plant species are classified as recalcitrant to regeneration by *in vitro* tissue culture, over the last two decades, enormous progress has been made in the application of biotechnological tools to the *Olea europaea*. The development of several protocols for micropropagation from axillary buds and somatic embryogenesis using explants isolated from selected adult trees has allowed the regeneration of several olive cultivars. However, the main challenge for mass production of olive plants by tissue culture is the production of plants that show genetic fidelity to the donor plant in a commercial production context. Fidelity is assured by the application of micropropagation by axillary buds, whereas somatic embryogenesis from long-term cultures does not guarantee genetic fidelity and leads to somaclonal variation among the regenerated plants.

Somaclonal variation in tissue culture is a complex problem that needs several approaches to be appreciated correctly. The use of only one type of molecular marker, such as RAPDs, to assess the genetic stability of an *in vitro* production system may be inadequate, and an approach that focuses on morphological traits appears to be a valuable complementary tool. For example, the analysis of morphological traits strongly confirms the specific and stable growth habit of regenerated plants by the analysis of vegetative progeny of somaclonal olive plants.

Future studies will be focused on target DNA fragments (AG1$_{550}$, OPA07$_{407}$, 1281$_{640}$ and OPP02$_{759}$) putatively associated with growth habit. These fragments were extracted from agarose gels and purified using the QIAquick Gel Extraction Kit (Qiagen), and the eluted putative growth habit-specific genes were cloned. Database searches for homologous sequences using the BLAST tool will be performed and, in addition, we are designing related oligonucleotide primers that will be synthesized and used to amplify sequence characterised amplified region (SCAR) markers.

The somaclonal variation generated by somatic embryogenesis presents a novel opportunity for olive breeders to experiment with new traits, in contrast to conventional long-term strategies for developing olive trees that have desirable new traits. A practical example of this potential is a dwarf olive tree identified among the BOS plants; the aesthetic, ecological and growth-habit characteristics of this individual support its use as an ornamental plant. Clones of this dwarf olive genotype could be planted in private gardens, public parks and on roadsides [137].

A.R. Leva produced the olive somaclones and has the exclusive rights on them.

R. Petruccelli and L.M.R. Rinaldi contributed to the morphological and molecular characterization of the olive somaclones.

Author details

A.R. Leva*, R. Petruccelli and L.M.R. Rinaldi
CNR IVALSA Trees and Timber Institute, Firenze, Italy

Acknowledgement

The authors wish to thank the Centro di Ricerca e Sperimentazione in Agricoltura (CRSA)"Basile Caramia" Locorotondo, Taranto, Italy and the University Department DPPMA and Vegetal Virology Institute CNR in Bari (Italy) for maintaining the daughter somaclonal olive plants.

9. References

[1]   Rani V, Raina S. Genetic fidelity of organized meristemderived micropropagated plants: a critical reappraisal. In Vitro Cellular Developmental Biology Plant 2000; 36: 319-330.

[2]   Rani V, Parida A, Raina S. Random amplified polymorphic DNA (RAPD) markers for genetic analysis in micropropagated plants of *Populus deltoides* Marsh. Plant Cell Report 1995; 14:459-462.

[3]   Steward FC. Growth and development of cultivated cells. III. Interpretation of the growth from free cell to carrot plant. American Journal of Botany 1958; 45:709-713.

---

* Corresponding Author

[4] Braun AC. A demonstration of the recovery of the crown-gall tumor cell with the use of complex tumors of single-cell origin. Proceedings of the National Academy of Sciences of the United States of America 1959; 45 : 932–938.

[5] Larkin P, Scowcroft W. Somaclonal variation a novel source of variability from cell cultures for plant improvement. Theoretical and Applied Genetics 1981; 60: 197-214.

[6] Karp A. Origins, causes and uses of variation in plant tissue cultures. In: Vasil IK, Thorpe TA (eds) Plant cell and tissue culture. Dordrecht : Kluwer Academic Publishers; 1994. p 139-152.

[7] Orbovic´V, Calovic´M, Viloria Z, Nielsen B, Gmitter F, Castle W, Grosser J. Analysis of genetic variability in various tissue culture-derived lemon plant populations using RAPD and flow cytometry. Euphytica 2008;161: 329-335.

[8] Skirvin RM, McPheeters KD, Norton M. Sources and frequency of somaclonal variation. HortScience 1994; 29:1232-1237.

[9] Kaeppler SM, Kaeppler HF, Rhee Y. Epigenetic aspects of somaclonal variation in plants. Plant Molecular Biology 2000; 43:179-188.

[10] Mehta YR, Angra DC. Somaclonal variation for disease resistance in wheat and production of dihaploids through wheat 9 maize hybrids. Genetics and Molecular Biology 2000; 23:617-622.

[11] Predieri S. Mutation induction and tissue culture in improving fruits. Plant Cell Tissue Organ Culture 2001; 64:185-210.

[12] Karp A. Somaclonal variation as a tool for crop improvement. Euphytica 1995; 85 295-302.

[13] Unai E, Iselen T, de Garcia E. Comparison of characteristics of bananas (*Musa sp.*) from the somaclone CIEN BTA-03 and its parental clone Williams. Fruit 2004; 59: 257-263.

[14] Skirvin RM, Norton M, McPheeters KD. Somaclonal variation:has it proved useful for plant improvement? Acta Horticulturae 1993; (336) 333–340.

[15] Skirvin RM, McPheeters KD, Norton M. Sources and frequency of somaclonal variation. HortScience 1994; 29 : 1232-1237.

[16] Pierik RLM. In vitro culture of higher plants. Kluwer Academic Publishers, Dordrecht; 1987.

[17] Skirvin RM Natural and induced variation in tissue culture. Euphytica 1978; 27: 241-266.

[18] Scowcroft WR. Genetic variability in tissue culture: impact on germplasm conservation and utilization. International board for plant genetic resources (IBPGR) technical report AGPGIBPGR/84/152, Rome; 1984.

[19] Sivanesan I. Shoot regeneration and somaclonal variation from leaf callus cultures of *Plumbago zeylanica* Linn. Asian Journal of Plant Science 2007; 6: 83–86.

[20] Vasil IK. Automation of plant propagation. Plant Cell Tissue Organ Culture 1994; 9:105-108.

[21] Ammirato PV, Styer DJ. Strategies for large scale manipulation of somatic embryos in suspension culture. In: Zaitlin M, Day P, Hollaender A. (eds.) Biotechnology in plant science: relevance to agriculture in the eighties. New York: Academic Press; 1985. 161- 178.

[22] Lutz JD, Wong JR, Rowe J, Tricoli DM, Lawrence RHJ. Somatic embryogenesis for mass cloning of crop plants. In: Henke RR, Hughes KW, Constantin MP, Hollaender A. (eds.)Tissue culture in forestry and agriculture. New York: Plenum; 1985. p105-116.

[23] Redenbaugh K, Viss P, Slade D, Fujii JA. Scale-up: artificial seeds. In: Green CE, Somers DA, Hackett WP, Biesboer DD. (eds.) Plant tissue and cell culture. New York: A. R. Liss; 1987. 473-493.

[24] Thorpe TA. Frontiers of plant tissue culture. Canada: University of Calgary Press; 1978.

[25] Thorpe TA. In vitro somatic embryogenesis. ISI atlas of science-animal and plant sciences. 1988; 1: 81-88.

[26] Debergh PC, Read PE. Micropropagation. In: Debergh PC, Zimmerman RH.(eds.) Micropropagation, technology and application. Dordrecht, Boston, London: Kluwer Academic; 1990. 1-13.

[27] Wang, P. J.; Charles, A. Micropropagation through meristem culture. In: Bajaj YPS. (ed.) Biotechnology in agriculture and forestry, HighTech and micropropagation I. New York: Springer Verlag 1991; (17) 32-52.

[28] Vasil IK. Somatic embryogenesis and its consequences in gramineae. In:Henke R, Hughes K, Constantin M, Hollaender A. (eds) Tissue culture in forestry and agriculture. New York: Plenum Press; 1985. p 31-47.

[29] Shenoy VB, Vasil IK. Biochemical and molecular analysis of plants derived from embryogenic cultures of napier grass (*Pennisetum purpureum* K. Schum.). Theoretical and Applied Genetics 1992; 83: 947-955.

[30] Murashige T.The impact of plant tissue culture on agriculture. In: Thorpe TA (ed.) Frontiers of plant tissue culture. Calgary: University of Calgary, International Association for Plant Tissue Culture; 1978. 15-26.

[31] Hammerschlag, FA. Factors influencing the frequency of callus formation among cultured peach anthers. HortScience 1984; 19:554.

[32] Gupta PK, Mascarenhas AF. Eucalyptus. In: Bonga JM, Durzan DJ (eds.) Cell and tissue culture in forestry. Dordrecht: Martinus Nijhoff; 1987. Vol. 3.p385-399.

[33] Eastman PAK, Webster FB, Pitel JA, Roberts DR. Evaluation of somaclonal variation during somatic embryogenesis of interior spruce (*Picea engelmannii* complex) using culture morphology and isozyme analysis. Plant Cell Report 1991; 10 :425–430.

[34] Isabel N, Tremblay L, Michaud M, Tremblay FM, Bousquet J. RAPDs as an aid to evaluate the genetic integrity of somatic embryogenesis derived populations of *Picea mariana* (Mill.) B.S.P. Theoretical and Applied Genetics 1993; 86 81-87.

[35] Gavidia I, Del Castillo ALD, Perez-Bermudez V. Selection of long-term cultures of high yielding *Digitalis obscura* plants: RAPD markers for analysis of genetic stability. Plant Science 1996; 121:197-205.

[36] Jain SM. Tissue culture-derived variation in crop improvement. Euphytica 2001; 118:153-166.

[37] Krikorian AD, Irizarry H, Cronauer-Mitra SS, Rivera E. Clonal fidelity and variation in plantain (*Musa* AAB) regenerated from vegetative stem and floral axis tips in vitro. Annals of Botany 1993; 71:519-535.

[38] Kawiak A, Łojkowska E. Application of RAPD in the determination of genetic fidelity in micropropagated Drosera plantlets. In Vitro Cellular Developmental Biology Plant 2004; 40:592-595.

[39] Chuang SJ, Chen CL, Chen JJ, Chou WY, Sung JM. Detection of somaclonal variation in micro-propagated Echinacea purpurea using AFLP marker. Scientia Horticulturae 2009; 120:121-126.

[40] Sahijram L, Soneji J, Bollamma K. Analyzing somaclonal variation in micropropagated bananas (*Musa spp.*). In Vitro Cellular Developmental Biology Plant 2003;39:551-556.

[41] Sharma S, Bryan G, Winfield M, Millam S. Stability of potato (*Solanum tuberosum* L.) plants regenerated via somatic embryos, axillary bud proliferated shoots, microtubers and true potato seeds: a comparative phenotypic, cytogenetic and molecular assessment. Planta 2007; 226:1449-1458.

[42] Kunitake H, Koreeda K, Mii M. Morphological and cytological characteristics of protoplast-derived plants of statice (*Limonium perezii* Hubbard). Scientia Horticulturae 1995; 60:305-312.

[43] Swartz H J. Post culture behaviour, genetic and epigenetic effects and related problems. In: Debergh PC, Zimmerman RH (eds.) Micropropagation: technology and application. Dodrecht : Kluwer Academic Publishers; 1991.p 95-122.

[44] Vidal M D C, De Garcìa E. Analysis of a *Musa spp.* Somaclonal variant resistant to yellow Sigatoka. Plant Molecular Biology Reporter 2000; 18: 23-31.

[45] Martin K, Pachathundikandi S, Zhang C, Slater A, Madassery J. RAPD analysis of a variant of banana (*Musa sp.*) cv. grande naine and its propagation via shoot tip culture. In Vitro Cellular Developmental Biology Plant 2006; 42:188-192.

[46] LoSchiavo F, Pitto L, Giuliano G, Torti G, Nuti-Ronchi V, Marazziti D, Vergara R, Orselli S, Terzi M. DNA methylation of embryogenic carrot cell cultures and its variations as caused by mutation, differentiation, hormones and hypomethylating drugs. Theoretical and Applied Genetics 1989; 77:325-331.

[47] Nehra NS, Kartha KK, Stushnott C, Giles KL. The influence of plant growth regulator concentrations and callus age on somaclonal variation in callus culture regenerants of strawberry. Plant Cell Tissue Organ Culture 1992; 29:257-268.

[48] Bouman H, De Klerk GJ. Measurement of the extent of somaclonal variation in begonia plants regenerated under various conditions. Comparison of three assays. Theoretical and Applied Genetics 2001;102: 111-117.

[49] Ahmed EU, Hayashi T, Yazawa S. Auxins increase the occurrence of leaf-colour variants in Caladium regenerated from leaf explants. Scientia Horticulturae 2004; 100:153-159.

[50] Mohanty S, Panda M, Subudhi E, Nayak S. Plant regeneration from callus culture of Curcuma aromatica and in vitro detection of somaclonal variation through cytophotometric analysis. Biologia Plantarum 2008; 52:783-786.

[51] Bayliss MW. Chromosomal variation in tissue culture. In:Vasil IK (ed.) Perspectives in plant cell and tissue culture. New York : Academic Press; 1980.p113-144.

[52] Venkatachalam L, Sreedhar RV, Bhagyalakshmi N. Genetic analyses of micropropagated and regenerated plantlets of banana as assessed by RAPD and ISSR markers. In Vitro Cellular Developmental Biology Plant 2007a; 43:267-274.

[53] Reuveni O, Israeli Y, Golubowicz S. Factors influencing the occurrence of somaclonal variations in micropropagated bananas. Acta Horticulturae 1993; 336:357-364.

[54] Giménez C, de Garcìa E, de Enrech NX, Blanca I. Somaclonal variation in banana: cytogenetic and molecular characterization of the somaclonal variant CIEN BTA-03. In Vitro Cellular Developmental Biology Plant 2001; 37:217-222.

[55] Roels S, Escalona M, Cejas I, Noceda C, Rodriguez R, Canal MJ, Sandoval J, Debergh P Optimization of plantain (Musa AAB) micropropagation by temporary immersion system. Plant Cell Tissue Organ Culture 2005; 82:57-66.

[56] Reuveni O, Israeli Y Measures to reduce somaclonal variation in in vitro propagated bananas. Acta Horticulturae 1990; 275:307-313.

[57] Rodrigues PHV, Tulmann Neto A, Cassieri Neto P, Mendes BMJ. Influence of the number of subcultures on somaclonal variation in micropropagated Nanico (*Musa spp.*, AAA group). Acta Horticulturae 1998; 490:469-473.

[58] Bairu MW, Fennell CW, van Staden J The effect of plant growth regulators on somaclonal variation in Cavendish banana (*Musa* AAA cv. 'Zelig'). Scientia Horticulturae 2006; 108:347-351.

[59] Israeli Y, Lahav E, Reuveni O. In vitro culture of bananas. In:Gowen S (ed.) Bananas and plantians. Chapman and Hall, London, 1995.p 147-178.

[60] Petolino JF, Roberts JL, Jayakumar P. Plant cell culture: a critical tool for agricultural biotechnology. In: Vinci VA, Parekh SR (eds.) Handbook of industrial cell culture: mammalian, microbial and plant cells. New Jersey : Humana Press; 2003.p 243-258.

[61] Cote F, Teisson C, Perrier X Somaclonal variation rate evolution in plant tissue culture: contribution to understanding through a statistical approach. In Vitro Cellular Developmental Biology Plant 2001; 37:539-542.

[62] Kaeppler S, Phillips R. DNA methylation and tissue culture induced variation in plants. In Vitro Cellular Developmental Biology Plant 1993; 29: 125-130.

[63] Shepherd K, Dos Santos JA Mitotic instability in banana varieties. I. Plants from callus and shoot tip cultures. Fruit 1996; 51:5-11.

[64] Popescu AN, Isac VS, Coman MS MSR Somaclonal variation in plants regenerated by organogenesis from callus cultures of strawberry (Fragaria 9 ananassa) Acta Horticulturae 1997; 439:89-96.

[65] Lee M, Phillips RL.The Chromosomal basis of somaclonal variation. Annual Review of Plant Physiology Plant Molecular Biology 1988; 39:413-437.

[66] Damasco OP, Smith MK, Adkins SW, Hetherington SE, Godwin ID. Identification and characterisation of dwarf off-types from micropropagated 'Cavendish' bananas. Acta Horticulturae 1998; 490:79-84.

[67] Stover RH. Somaclonal variation in Grande Naine and Saba bananas in the nursery and in the field. In: Persley GJ, De Langhe E (eds.) ACIAR proceeding no. 21, Canberra; 1987.

[68] Israeli Y, Reuveni O, Lahav E. Qualitative aspects of somaclonal variations in banana propagated by in vitro techniques. Scientia Horticulturae 1991; 48:71-88.

[69] Martin K, Pachathundikandi S, Zhang C, Slater A, Madassery J. RAPD analysis of a variant of banana (*Musa sp.*) cv. grande naine and its propagation via shoot tip culture. Plant 2006; 42:188-192.

[70] D'Amato F. Cytogenetics of differentiation in tissue and cell culture. In: Reinert J, Bajaj YPS (eds.) Applied and fundamental aspects of plant cell, tissue and organ culture. New York : Springer; 1977, p 343-464.

[71] Hao Y-J, Deng X-X. Occurrence of chromosomal variations and plant regeneration from long-term-cultured *citrus* callus. In Vitro Cellular Developmental Biology Plant 2002; 38:472-476.

[72] Mujib A, Banerjee S, Dev Ghosh P. Callus induction, somatic embryogenesis and chromosomal instability in tissue culture raised hippeastrum (*Hippeastrum hybridum* cv. United Nations). Propagation of Ornamental Plants 2007; 7:169-174.

[73] Hirochika H Activation of tobacco transposons during tissue culture. EMBO J 1993; 12:2521-2528.

[74] Barret P, Brinkman M, Beckert M. A sequence related to rice Pong transposable element displays transcriptional activation by in vitro culture and reveals somaclonal variations in maize. Genome 2006; 49:1399–1407.

[75] Smulders MJM, de Klerk G J. Epigenetics in plant tissue culture. Plant Growth Regulation 2011; 63:137-146

[76] Jaligot E, Rival A, Beule' T, Dussert S, Verdeil JL Somaclonal variation in oil palm (*Elaeis guineensis* Jacq.): the DNA methylation hypothesis. Plant Cell Report 2000; 19:684-690.

[77] Schellenbaum P, Mohle V, Wenzel G, Walter B. Variation in DNA methylation patterns of grapevine somaclones (*Vitis vinifera L.*) BMC Plant Biology 2008; 8:78-88.

[78] Baránek M., Křižan B., Ondružıková E., Pidra M. DNA methylation changes in grapevine somaclones following in vitro culture and thermotherapy. Plant Cell Tissue Organ Culture 2010; 101:11-22.

[79] Li X, Xu M, Korban S S. DNA methylation profiles differ between field- and in vitro-grown leaves of apple. Journal of Plant Physiology 2002; 159: 1129-1134.

[80] Hammerschlag FA. Somaclonal variation. In: Hammerschlag FA Litz RE (eds)CAB International Biotechnology of Perennial Fruit Crops. Wallingfors, 1992. p 35-36.

[81] Karp, A. On the current understanding of somaclonal variation. In: Miflin B J (ed.), Surveys of Plant Molecular and Cell Biology, Oxford University Press; 1982.7: 1-58.

[82] Piagnani M C, Maffi D, Rossoni M, Chiozzotto R. Morphological and physiological behaviour of sweet cherry 'somaclone' HS plants in field. Euphytica 2008; 160:165-173.

[83] Faure O, Nougarède A. Nuclear DNA content of somatic and zygotic embryos of *Vitis vinifera* cv. Grenache Noir at the torpedo stage-flow cytometry and in situ DNA microspectrophotometry. Protoplasma 1993; 176:145-150.

[84] Kuksova VB, Piven NM, Gleba YY Somaclonal variation and in vitro induced mutagenesis in grapevine. Plant Cell Tissue Organ Culture 1997; 49:17-27.

[85] Leal F, Loureiro J, Rodriguez E, Pais MS, Santos C,·Pinto-Carnide O. Nuclear DNA content of *Vitis vinifera* cultivars and ploidy level analyses of somatic embryo-derived plants obtained from anther culture. Plant Cell Report 2006; 25: 978-985.

[86] Carini F, De Pasquale F. Micropropagation of *Citrus*. In Micropropagation of Wood Tree and Fruits. S Mohan and katsuaki Ishii (ed.). London : Kluwer Academic Publishers ; 2003; 75:589-619.

[87] Fourré JL Somaclonal variation and genetic molecular markers in woody plants. In: Jain SM, Minocha SC (eds.) Molecular biology of woody plants. The Netherlands : Kluwer; 2000. p. 425-449

[88] Veilleux RE, Johnson AAT. Somaclonal variation: Molecular analysis, transformation, interaction, and utilization. Plant Breeding Reviews 1998;16: 229-268.

[89] Hashmi G, Huettel R, Meyer R, Krusberg L, Hammerschlag F. RAPD analysis of somaclonal variants derived from embryo callus cultures of peach. Plant Cell Report 1997; 16:624–627.

[90] Piola F, Rohr R, Heizmann P. Rapid detection of genetic variation within and among in vitro propagated cedar (*Cedrus libani* Loudon) clones. Plant Science 1999; 141:159-163.

[91] Renau-Morata B, Nebauer SG, Arrillaga I, Segura J. Assessments of somaclonal variation in micropropagated shoots of cedrus: consequences of axillary bud breaking. Tree Genetics and Genomes 2005; 1: 3–10.

[92] Khan EU, Fu XZ, Wang J, Fan QJ, Huang XS, Zhang GN, Shi J, Liu JH Regeneration and characterization of plants derived from leaf in vitro culture of two sweet orange (*Citrus sinensis* (L.) Osbeck) cultivars. Scientia Horticulturae 2009; 120:70-76.

[93] Oliveira CM, Mota M, Monte-Corvo L, Goulão L, Silva DM. Molecular typing of Pyrus based on RAPD markers. Scientia Horticulturae 1999; 79: 163–174.

[94] Nas M N, Multu N, Read PE. Random amplified polimorfic DNA (RAPD) analysis. HortScience 2004; 39:1079-1082.

[95] Virscek-Marn M, Bohanec B, Javornik B. (1999) Adventitious shoot regeneration from apple leaves optimisation of the protocol and assessment of genetic variation among regenerants. Phyton 39: 61-70.

[96] Caboni E, Lauri P, Damiano C, D'Angeli S. Somaclonal variation induced by adventitious shoot regeneration in pear and apple. Acta Horticulture 2000; 530: 195-202.

[97] Carvalho LC, Goulao L, Oliveira C, Gonçalves JC, Amancio S. RAPD assessment for identification of clonal identity and genetic stability of *in vitro* propagated chestnut hybrids. Plant Cell Tissue Organ Culture 2004;77: 23-27.

[98] Prado MJ, Herrera MT, Vázquez RA, Romo S, González MV. Micropropagation of two selected male kiwifruit and analysis of genetic variation with AFLP markers. HortScience 2005; 40: 740-746.

[99] Prado MJ, Gonzalez MV, Romo S, Herrera MT. Adventitious plant regeneration on leaf explants from adult male kiwifruit and AFLP analysis of genetic variation. Plant Cell Tissue Organ Culture 2007; 88: 1-10.

[100] Popescu C F, Flak A, Glimelius K. Application of AFLPs to characterize somaclonal variation in anther-derived grapevines. Vitis 2002; 41:177-182.

[101] Xiangqian Li, Mingliang Xu, Schuyler S. Korban DNA methylation profiles differ between field- and *in vitro*-grown leaves of apple. Journal Plant Physiology 2002; 159: 1229-1234.

[102] Martinelli L, Zambanini J, Grando MS. Genotype assessment of grape regenerants from floral explants. Vitis 2004; 43:119-122.

[103] Nookaraju A, Agrawal DC. Genetic homogeneity of in vitro raised plants of grapevine cv. Crimson Seedless revealed by ISSR and microsatellite markers. *South African Journal of Botany* 2012; 78: 302-306.

[104] Pathak H, Dhawan V. ISSR assay for ascertaining genetic fidelity of micropropagated plants of apple rootstock Merton 793.In Vitro Cellular Developmental Biology Plant 2012; 48: 137-143.

[105] Palombi MA, Damiano C. Comparison between RAPD and SSR molecular markers in detecting genetic variation in kiwifruit (*Actinidia deliciosa* A Chev). Plant Cell Rep 2002; 20: 1061–1066.

[106] Sarmento D, Martins M, Oliveira MM. Evaluation of somaclonal variation in almond using RAPD and ISSR. Options Méditerranéennes, Série A 2005; 63: 391-395.

[107] Gribaudo I, Torello Marinoni D, Gambino G, Mannini F, Akkak A, Botta R. Assessment of genetic fidelity in regenerants from two *Vitis vinifera* cultivars Acta Horticulturae 2009; 827:131-136.

[108] Bonga JM, Klimaszewska KK, von Aderkas P. Recalcitrance in clonal propagation, in particular of conifer. Plant Cell Tissue Organ Culture 2010; 100:241–254.

[109] Schaeffer WI. Terminology associated with cell, tissue and organ culture, molecular biology and molecular genetics. In Vitro Cellular Developmental Biology Plant 1990; 26: 97–101.

[110] Grossoni P. Problems concerning the in vitro culture of *Olea europaea* L. Giornale Botanico Italiano 1979; 113:75-88.

[111] Rugini E. *In vitro* propagation of some (*Olea europaea* L.) cultivars with different root-ability and medium development using analytical data from developing shoots and embryos. Scientia Horticulturae 1984; 24: 123-134.

[112] Fiorino P, Leva AR. Investigations on the micropropagation of the olive (*Olea europaea* L.) and influence of some mineral elements on the proliferation and rooting of explants. Olea 1986; 17: 101-104.

[113] Leva AR, Petruccelli R, Muleo R, Goretti R, Bartolini G. Influenza di fattori trofici, regolativi e condizioni di coltura in vitro di diverse cultivar di olivo. Atti Convegno "L'Olivicoltura Mediterranea: Stato e Prospettive della Coltura e della Ricerca. Rende 26–28 Gennaio 1995. p239-248.

[114] Grigoriadou K, Vasilakakis M, Eleftheriou EP *In vitro* propagation of the Greek olive cultivar "Chondrolia Chalkidikis". Plant Cell Tissue Organ Culture 2002; 71: 47-54.

[115] Leva AR, Sadeghi H, Petruccelli R. Carbohydrates Modulate the In Vitro Growth of Olive Microshoots. I. The Analysis of Shoot Growth and Branching Patterns. Journal of Plant Growth Regulation 2012. DOI 10.1007/s00344-012-9275-7

[116] Leva AR. Innovative protocol for "ex vitro rooting" on olive micropropagation Central European Journal of Biology. 2011; 6(3) : 352-358.

[117] Rugini E. Somatic embryogenesis and plant regeneration in Olive (*Olea europaea* L.). Plant Cell Tissue and Organ Culture 1988; 14: 207-214.

[118] Leva AR, Muleo R, Petruccelli R Long-term somatic embryogenesis from immature olive cotyledons. Journal Horticultural Science (1995b); 70: 417-421.

[119] Rugini E, Caricato G. Somatic embryogenesis and recovery from mature tissues of olive cultivars (*Olea europaea* L.) "Canino" and "Moraiolo". Plant Cell Report 1995; 14 : 257–260.

[120] Peyvandi M, Dadashian A, Ebrahimzadeh A, Majd A. Embryogenesis and rhizogenesis in mature zygotic embryos of olive (*Olea europaea* L.) cultivars *Mission* and *Kroneiki*. Journal of Sciences, Islamic Republic of Iran 2001; 12(1):9–15.

[121] Trabelsi EB, Bouzid S, Bouzid M, EIIoumi N, Belfeleh Z, Benabdallah A, and Ghezel R In Vitro Regeneration of Olive Tree by Somatic Embryogenesis Journal of Plant Biology 2003; 46: 173-180.

[122] Peyvandi M, Nemat Farahzadi H, Arbabian S, Noormohammadi Z, Hosseini-Mazinani M. Somaclonal Variation Among Somatic-Embryo Derived Plants of *Olea europaea L "cv. Kroneiki".* Journal of Sciences, Islamic Republic of Iran 2010; 21 (1): 7-14.

[123] Capelo AM, Silvia S, Brito G, Santos C. Somatic embryogenesis induction in leaves and petioles of a mature wild olive. Plant Cell Tissue Organ Culture 2010; 103:237-242.

[124] Mazri MA, Elbakkali A, Belkoura M, Belkoura I. Embryogenic competence of calli and embryos regeneration from various explants of Dahbia cv, a moroccan olive tree (*Olea europaea* L.). African Journal of Biotechnology 2011; 10 (82):19089-19095.

[125] Trabelsi EB, Naija S, Elloumi N, Belfeleh Z, Msellem M, Ghezel R, Bouzid S. Somatic embryogenesis in cell suspension cultures of olive *Olea europaea* L. 'Chetoui' Acta Physiologiae Plantarum 2011; 33:319-324

[126] Briccoli Bati C, Godini G, Nuzzo V. Preliminar agronomic evaluation of two cultivars of olive trees obtained from micropropagation method. Acta Horticulture 2002; 586: 867-870.

[127] Leva AR, Petruccelli R, Montagni G, Muleo R. Field performance of micropropagated olive plants (cv Maurino): morphological and molecular features. Acta Horticulture 2002; 586: 891-894.

[128] Leva AR. Morphological evaluation of olive plants propagated *in vitro* through axillary buds and somatic embryogenesis methods. African Journal of Plant Science 2009; 3 (3): 37-43.

[129] Roselli G, Mariotti P, Tessa A. Caratterizzazione di progenie di olivo mediante analisi chimica e sensoriale degli olii. Proceedings del Simposio Nazionale : "Germoplasma olivicolo e tipicità dell'olio" tenuto a Perugia 5 dicembre, 2003. pp 278-283

[130] Barranco D, Rallo L. Olive cultivars in Spain. HorTechnology 2000; 10:107–110.

[131] Rotondi A, Magli M, Ricciolini C, Baldoni L. Morphological and molecular analyses for the characterization of a group of Italian olive cultivars. Euphytica 2003; 132: 129–137.

[132] Leva AR, Petruccelli R, Polsinelli L. In vitro olive propagation: from the laboratory to the production line. OLIVAE 2004; 101: 18–26.

[133] Gucci R, Cantini C. Pruning and Training Systems for Modern Olive Growing. CSIRO Publishing, Australia 2000.

[134] Leva AR, Petruccelli R. Monitoring of cultivar identity in micropropagated olive plants using RAPD and ISSR markers. Biologia Plantarum 2012; 56 (2): 373-376.

[135] Mencuccini M.Produzione di somacloni da cultivar di olivo e prime osservazioni in campo Acta Italus Hortus 2011; 1: 113-116.

[136] Leva AR, Muleo R, Petruccelli R. Stabilità in campo di diversi habitus vegetativi in individui di *Olea europaea* L (cv Frangivento) derivati da embriogenesi somatica.Proceedings of the National Congress : Biodiversità-Opportunità di Sviluppo Sostenibile. Istituto Agronomico Mediterraneo, Bari, Italy. 2001.Volume II. 407-414.

[137] Leva AR, Petruccelli R. Dwarf olive trees for ornamental use: a morphological evaluation Journal of Horticultural Science & Biotechnology 2011; 86 (3) 217–220.

Chapter 7

Flow Cytometry Applied in Tissue Culture

Moacir Pasqual, Leila Aparecida Salles Pio,
Ana Catarina Lima Oliveira and Joyce Dória Rodrigues Soares

Additional information is available at the end of the chapter

1. Introduction

Flow cytometry is a powerful technology that allows for the simultaneous analysis of multiple attributes of cells or particles in a liquid medium. The first cytometer used was built during World War II, when [1] developed an equipment where particles flowed through the system to diffuse light through a lens, producing electrical signals sensed by a photodetector. The instrument could detect objects in the order of ~ 0.5 µm in diameter, and is recognized as the first flow cytometer used for observation of biological cells [2]. This would be possible to identify aerosols, bacteria that would possibly biological warfare agents as well as check the efficiency of gas mask filters against particles. In 1950, the same principle was applied to the detection and enumeration of blood cells. As hematology and cellular immunology, two biological areas, that drove the development of flow cytometry [3]. Later, with improved equipment and methods, this technique was adapted to other areas of biology, including the plant kingdom [4]. Already in 1973 the German botanist Friedrich Otto Heller used the Impulszytophotometrie (pulse cytophotometry in German). This scientist did not imagine that it has launched a new field of scientific research, which would later be called flow cytometry in plants.

In reference to [5] that developed a rapid and convenient method for the isolation of plant nuclei by cutting the same tissue in a lysis buffer consisting of a buffer to destroy the cellular and nuclear membranes of the cell allowing the release of DNA. Since then, this has been the main and most reliable method of isolating nuclear plant in flow cytometry. Any type of sample can be analyzed because its particles (cells, nuclei, chromosomes, cell organelles, or other cell subparticles) are suspended and vary between 0.2 µm and 50 µm in size. Solid tissues must be disaggregated and suspended before flow cytometry analysis. The suspended particles are then placed into a flow cytometry device.

The studies on flow cytometry have used as base the plant tissue culture, including the regeneration of plants subjected to chromosome doubling, for detection of somaclonal variation in material micropropagated in various subcultures, viability of pollen grains, cell cycles and the determination of ploidy. This chapter presents results obtained through flow cytometry on plant tissue culture.

2. Preparation of material for analysis in flow cytometry

There are several methods that can be used to prepare plant material and to estimate the DNA content by flow cytometry. The methodologies differ according to plant species, a laboratory, with the brand and model flow cytometer used. In Tissue Culture Lab in the DAG / UFLA the methodology used is described in Figure 1.

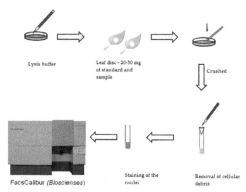

Figure 1. Diagram of the methodology used to analyze the nuclear DNA content from plant tissue. Source: Adapted from [6].

The sample must be in the form of a suspension of single particles [7]f or being analyzed by flow cytometry. Analysis of DNA content by flow cytometry is based on the fluorescence intensity of nuclei stained with a fluorochrome specific to the DNA. There is problem related with low capacity of penetration of fluorochromes but this can be overcomed if the nuclei are released prior to staining. [7]. Secondary metabolites can interfere in cellular content and color of the fluorescent dye [8].

There are several methodologies developed for the release of the nuclei of plant tissues. However, the methodology proposed in reference [9] can promote the release of the nucleus (Figure 1) and is frequently used for simplicity and speed. The differences observed between the methods are the composition of the lysis buffer for isolation of nuclei, the fluorochrome for nuclear staining of the suspension and reading the sample in the flow cytometer.

Plant tissue samples are perforated, with the aid of a cutting blade, in a buffer solution for the extraction and isolation of nuclei. Subsequently, the suspension is filtered by a fine mesh nylon (20-100 µM pore diameter) [10]. This filtering is performed in order to remove all the material in the sample greater than the core, leaving in solution only those estimates and thereby obtaining the DNA content of more reliable. Furthermore, the presence of other components and soluble substances such as chloroplasts, mitochondria, phenolic compounds, DNAse, RNAse etc., which are released in the cytosol may through this filter and compromise the quality of results. An alternative may be employed to remove such debris is the washing of the nuclei using centrifugation and resuspension, and to modify the components and / or pH of the buffer [11]. After filtering the samples are stained with a fluorochrome specific, then the analysis of samples in the flow cytometer.

3. Factors that affect the quality of the sample

Several factors can affect the quality of the samples and consequently the reliability of estimates of DNA content obtained by flow cytometry. Extraction buffer, reference standard, fluorochrome, type of plant tissue used (chemical composition and the presence of anthocyanin, phenolic compounds that inhibit DNA staining), quality of the sample (plant age, presence of injuries, diseases ...), storage time of the plant tissue, care in preparation and sample analysis are among the factors involved [12]. Thus, an appropriate methodology is necessary for each species.

3.1. Nuclear extraction buffer

The extraction buffer is an appropriate solution that has the function to release the nuclei of intact cells, preserving and ensuring the stability and integrity of nuclei during the experiment, inhibiting the activity of nucleases, and providing optimal conditions for staining of DNA by stoichiometry [13]. Approximately 25 caps are, but only eight are commonly used in flow cytometry [14]. The six most commonly used buffers are shown in Table 1.

Buffer	Composition	Standard
Galbraith	45 mM MgCl2; 30 mM citrato de sódio; 20 mM MOPS; 0.1% (v/v) Triton X-100; pH 7.0	[15]
LB01[1]	15 mM Tris; 2 mM Na2 EDTA; 0.5 mM espermina.4HCl; 80 mM KCl; 20 mM NaCl; 0.1% (v/v) Triton X-100; pH 7.5	[16]
Otto's	Otto I: 100 mM ácido citric mono hidratado; 0.5% (v/v) Tween 20 (pH approx. 2–3) Otto II: 400 mM Na2PO4.12H2O (pH approx. 8–9)	[17], [18]
Tris.MgCl2	200 mM Tris; 4 mM MgCl2.6H2O; 0.5% (v/v) Triton X- 100; pH 7.5	[19]
Marie	50 mM glucose; 15 mM NaCl; 15 mM KCl; 5 mM Na2 EDTA; 50 mM citrate de sódio; 0.5% Tween 20, 50 mM HEPES (pH 7.2), 1% (m/v) polyvinylpyrrolidone-10 (PVP-10)	[20]

EDTA = ethylenediamine tetraacetic acid; HEPES = 4-2 ethanesulfonic Acid Hydroxyethyl-piperazine-1; MOPS = 3 - (N-morpholino) propanesulfonic; = Tris (hidroximetril) aminomethane and PVP = polyvinyl pyrrolidone.

Table 1. Composition of extraction buffers commonly used in flow cytometry plant.

The caps have in their composition organic buffering substances, non-ionic detergents and stabilizers of chromatin. The substances commonly used are buffers, MOPS, HEPES, and

TRIS, allowing the stabilization of pH 7-8 solutions, which is the pH range compatible to most of the fluorochromes used.

The nonionic detergents are present in the buffer solutions with TRITON X-100 and Tween-20, for cleaning of the cores and separation for avoiding that they add to each other or with possible debris present in the sample.

Stabilizers used in the composition of the buffer are $MgCl_2$, $MgSO_4$ and spermine and chelating agents such as EDTA and sodium citrate. These components bind divalent cations which are cofactors endonuclease. The inorganic salts NaCl and KCl allow to achieve adequate ionic strength [21].

Cytosolic compounds that are released during the isolation of nuclei, interact with nuclear DNA and / or the fluorochrome, and affect the quality of the sample and cause stoichiometric errors [22, 23, 24, 25].

In the literature there are few reports that compare the efficiency of different buffers for nuclear extraction. There is a single buffer works optimally for all types or tissues and plant species, previous studies are needed to identify the most appropriate buffer for each species studied and contribute to a greater experimental precision [24].

3.2. Reference standards fluorochromes

The reference standard is a DNA of species whose amount already previously known, and thus can be estimated by comparing the DNA content of any kind. There are a number of reference patterns with a wide range of DNA content allowing coverage of a wide range of genome. a species whose amount The use of these standards allows comparison of results obtained in different laboratories.

Estimates of DNA content obtained by flow cytometry are always relative to a standard whose DNA content is already established. This pattern receives two reference designations internal standard, when extraction of the cores and the analysis of sample and standard are performed simultaneously, or when an external standard is performed separately. The internal standards are most recommended, especially in high-precision measurements, because the peaks of the standard used and the sample appear in the same histogram and are treated under identical conditions [26] thereby reducing possible errors due to oscillation of the device during the evaluation of the samples. However [27] reported that the simultaneous processing of the sample and the reference standard was not necessary to obtain reliable estimates of DNA. It is common to use only one reference standard in all analyzes of the same experiment, but this procedure carries the risk of error due to nonlinearity [28, 29].

However, the choice and correct use of reference standards is a criterion that has been largely neglected [30].

The researcher Jaroslav Doležel from Laboratory of Molecular Cytogenetics and Cytometry, of the Czech Republic has set benchmarks with content from genomic DNA with different sizes.

Description	DNA content (pg)	References
Raphanus sativus cv Saxas	1,11	[31]
Solanum lycopersicum cv Stupické	1,96	[21]
Glycine max	2,5	
Zea mays	5,72	
Pisum sativum cv Ctirad	9,09	[32]
Secale cereale	16,19	
Vicia faba	26,90	
Allium cepa	34,89	

Table 2. Content of DNA of known standards are used.

3.3. Fluorochromes

The choice of fluorochrome is another important factor that affects the reliability of estimates of DNA content. The fluorochromes specifically bind to DNA and stoichiometrically in accordance with the intensity of fluorescence of the nucleus or the cell suspensions analyzed on flow is estimated for DNA content [6]. Fluorochromes used in coloring cores are shown in Table 3.

Fluorochrome	DNA binding mode	Wave-length	
		Excitation	Emission
Propidium iodide	Interleaving	525 (Blue-green)	605 (Red)
Ethidium bromide	Interleaving	535 (Blue-green)	602 (Red)
SYBR Green	Interleaving	488 (Blue)	522 (Green)
DAPI	Rich regions in AT	345 (UV)	460 (Blue)
Hoechst 33258	Rich regions in AT	360 (UV)	460 (Blue)
Chromomycin A3	Rich regions in GC	445 (Violet-blue)	520 (Green)
Mithramycin	Rich regions in GC	445 (Violet-blue)	575 (Green)

Table 3. Fluorochromes used in flow cytometry to estimate the DNA content.

There are two classes of fluorochromes the intercalating and specific. The propidium iodide, ethidium bromide and Sybr Green are intercalating fluorochrome, i.e., without preference of base pairs and are the most adequate to estimate the DNA content [33 cited by 34].

DAPI, Hoechst 33258, Chromomycin A3 and Mithramycin fluorochromes are specific. The Mithramycin, Chromomycin and the Olivomycins are fluorochromes which preferentially bind regions of DNA in GC-rich [35]. While the fluorochrome DAPI and Hoechst (33342 and 33258) were also specific DNA binds to AT-rich regions [36]. Therefore, the use of these dyes can lead to many incorrect estimates of the values of DNA content, since it is not known in advance the ratio of AT GC in species to be estimated the DNA content.

The propidium iodide has the lowest coefficient of variation obtained in using the fluorochrome is most suitable for determining the amount of genomic DNA in plants [37, 38, 39]. However, other authors reported propidium iodide and ethidium bromide are not dye specifically the DNA, they dye RNA too, but to not compromise the efficiency of the determination of content DNA can be used RNase [40].

4. Care use of cytometry

Below are listed some precautions that should be taken during the use of flow cytometry:

1. Avoid filling the tank of saline to their maximum capacity. When a tank is filled with pressurized fluid is forced toward the air hose preventing adequate pressurisation of the enclosure.
2. When working with propidium iodide, should be placed approximately 400 ml of hypochlorite in the sewage tank, which has a capacity of 4 liters, since the chlorine inactive molecules iodide.
3. It should be cleaned daily after use of the cytometer, the following steps: with the "RUN" button, install the probe tubes containing 3 ml of 0.5% hypochlorite, left to run on HI for 1 minute with the arm 5 minutes to open and close the arm. Select the fluid control "STNDBY." Remove the tube and insert another tube containing 1 ml rince facs (which is a detergent that helps remove waste from dyes into the machine) and let it run for 2 minutes in HI, with the arm closed. Select the button again fluid control "STNDBY" Remove everything and place another tube containing 3 ml of distilled water and let it run one minute with the open arm in HI and 5 minutes with the arm closed. Select button "STANDBY" and then install a tube containing no more than 1 ml of distilled water in the probe, because it always returns to the saline and the tube makes the volume of the tube exceeds its maximum capacity if it has more than 1 ml of distilled water, and this can affect equipment performance.
4. The tube should remain in distilled water to prevent probe salt deposits are formed in the sample injection tube
5. It should be cleaned monthly. This procedure is performed on the entire fluid system and once a month, or more often as needed. It should be removed from the reservoir containing saline solution and then install a different container with 1-2 liters of 0.5% sodium hypochlorite, flush for 30 minutes, while in the probe set 3 ml of hypochlorite solution at the same concentration . After this period must be installed to another container containing 1 to 2 liters of distilled water and left to run for 30 minutes, while the probe install a tube containing 3 ml of distilled water at the same concentration. During this procedure, iodide should never pass through the filter of saline, as you may damage it, so the hose to the filter should be disconnected during this process. Following the procedure returns the brine tank to the right place and connect the hose from the filter.
6. If the equipment becomes more than a week without being used, the salt tank must be replaced by distilled water and left to run for about 10 minutes to remove any salt of the capillary tubes of the equipment, because the salt form crystals which can clog the entire system.

7. Never replace the air tube into the sample if the button "HI" is on, the tubes should always be replaced with the "STNDBY" button and you must not allow the sample to be sucked through the probe, thus preventing air from fluid system.

8. All bubbles are displayed in the hoses from the tank and filter salt must be removed before the reading of the samples, because it makes the reading very slow. If you suspect bubbles within the system must press the "PRIME", because it injects a blast of air across the system and then complete with saline, removing bubbles. This procedure should be repeated 5 times to really solve the problem.

9. Should perform preventive maintenance on a flow cytometer, once a year by specialized professionals.

5. How to troubleshoot an analysis of flow cytometry

Paul Kron of Integrative Biology University of Guelph 10 list of solutions to problems have a histogram of quality estimates DNA content trusted. These solutions are listed below and have some adjustments based on the experiences gained at work in the Laboratory of Tissue Culture UFLA.

5.1. Verify that the flow cytometer is running well and is configured correctly

A quality control test should be performed daily and periodic maintenance by a technician from the manufacturer. These precautions ensure the proper functioning of the device.

Verify that the parameters were set by someone who is qualified to do so. Depending on the application we can use fluorescence intensity (height) or integrated fluorescence (area), linear or logarithmic scale and is vital to know the parameter most suitable for your dye.

5.2. Use good quality plant tissues

For most samples sheets are used, which should be healthy, young and cool. Sheet that shows any sign of senescence should be avoided; leaf collected at the end of the growing season often does not work. Avoid using wilted leaves.

For some species the leaves can be stored in refrigerator for 1 to 5 days after collection, since it kept in sealed plastic bag with some moist cotton. Do not leave the sample in direct contact with ice, or excessive moisture. It is also possible to store dried tissue, making use of desiccants substances. More tests are needed in this area to define protocols desiccation.

5.3. Use the appropriate tissue

If the swatch does not work, it is possible to test embryos, shoots, roots, flower petals, fruit or other healthy tissue. However, for certain species may occur the endopolyploidy, i.e. the degree of ploidy may vary between tissues, several peaks appearing in the histogram. In this case must be used whenever the first peak to DNA content.

In case of use of seeds is necessary to attend the endosperm and embryo differ in ploidy, and the seeds may be hybrids [41].

5.4. Use the correct buffer

The choice of buffer can have a huge impact on the quality of data. This choice can influence the relative fluorescence, and the quality peak [11].

It is necessary to test not only buffers, but also the consistency of results. It is possible that a buffer can lead to production of very clean samples with low CV, but in highly variable repetition of the measures of fluorescence [42].

The pH of the buffer must be between pH 7-8.

5.5. Ajuste the quantity of tissue and / or excessive cutting the sample

Excess sample is cut on a common problem and can overload the buffer, reducing their ability to maintain the correct pH range, dark coloration and large amounts of precipitation are not good signs. Keep samples on ice during cutting may help. It is possible to improve the quality of the sample cut by at least increase the amount of buffer, or by reducing the amount of tissue in the sample.

It is important to worry about getting good quality at the peaks (low CV), not number of cores. One should not impair the quality of the sample in search of "10.000 colors." This approach is often misapplied, and is more usual in analyzes of cell cycle. The core guide 1300 is the best for many applications [42]. A clean sample of 500 events per peak will probably tell you more than 10.000 events with peaks of large particles and high CV histogram very jagged.

5.6. Adjust the conditions of time and coloration.

After 2 hours of sample preparation buffer, the cores may begin to degrade. Ideally, the sample should be read in a short time after staining with 10 minutes to 2 hours, as the extreme limits.

During the stages of sample preparation, staining and reading is essential to keep them on ice and then the color should keep them in the dark, not to lose fluorescence until the moment you put them on the cytometer.

5.7. Try centrifugation

An alternative to improve the quality of the histograms is cut into a sample buffer, centrifuged (slow speed for 05-10 min), remove the supernatant and suspended again the pellet in 0.5 ml buffer, then filter and staining. This can clean up some samples.

5.8. Try a different pattern

Histogram bad when you are on a second species such as an internal standard, there may be interference between the two species of plants used (for example, by the effect of secondary metabolites) [24].

5.9. Make a gate in their samples

Even when the peaks are small and there is debris (dirt), the peaks can be measured with appropriate software making Gates. However, the removal of debris through the gate can affect how the curve fitting software analyzes of the histograms. Moreover, by making a very large suppression of scattering nuclei generates peak with a CV that both subjective and possibly artificially low, so methods of gate should be clearly described in any publication.

The samples with large amounts of debris over the cores must be considered suspect because the debris may be interfering in the coloring. Gate histograms in such poor quality must be made only when all other options fail.

Some other things to consider:

- Some tissue types may require special approaches. For example, pollen cores can be difficult to extract, as well as cutting methods and may be required for a review, see [43].
- Not all flow cytometers are equal. Some may produce better results than others, depending on factors such as size of the nuclei. If you have the opportunity to try more than one machine, the results can be enlightening.

6. Applications of plant flow cytometry

6.1. Tissue culture

Flow cytometry and microsatellite analyses were used to evaluate the trueness-to-type of somatic embryogenesis-regenerated plants from six important Spanish grapevine (*Vitis vinifera* L.) cultivars. Tetraploid plants were regenerated through somatic embryogenesis from all of the cultivars tested with the exception of 'Merenzao'. In addition, an octoploid plant was obtained in the cv. 'Albariño', and two mixoploids in 'Torronte's'. The most probable origin of these ploidy variations is somaclonal variation. The cv. 'Brancellao' presented significantly more polyploids (28.57%) than any other cultivar, but it must be noted that 50% of the adult field-grown 'Brancellao' mother plants analysed were mixoploid. Hence, it is probable that these polyploids originated either from somaclonal variation or by separation of genotypically different cell layers through somatic embryogenesis. Microsatellite analysis of somatic embryogenesis-regenerated plants showed true-totype varietal genotypes for all plants except six 'Torronte's' plants, which showed a mutant allele (231) instead of the normal one (237) at the locusVVMD5. There was not a clear relationship between the occurrence of the observed mutant regenerated plants and the callus induction media composition, the developmental stage of the inflorescences, the type of explant used for starting the cultures or the type of germination (precocious in differentiation medium or normal in germination medium) in any of the cultivars tested, except 'Torronte's' [44].

In addition, flow cytometry was used in breeding programmes to determine ploidy status after colchicine treatment of banana plants.

In reference [45] objective was to assess the colchicine and amiprophos-methyl (APM) concentration and exposure period in the chromosome duplication of banana plants diploids. Banana stem tips were used from the following genotypes: breed diploids (1304-04 [Malaccensis x Madang (*Musa acuminata* spp. Banksii)] and 8694-15 [0337-02 (Calcutta x Galeo) x SH32-63]). Colchicine was used at concentrations of 0 (control treatment), 1.25, 2.5 and 5.0 mM, while APM was used at 0 (control treatment), 40 and 80 μM, in solution under agitation (20 rpm), for 24 and 48 h periods. With the use of APM, 66.67% tetraploid plants were obtained in the 1304-04 genotype using 40 μM for 24 h and 18.18% in 80 μM for 48 h, while in the 8694-15 genotype using 40 and 80 μM colchicine for 48 h, 27.27 and 21.43% tetraploid plants were observed, respectively. For colchicine, in the 1304-04 genotype, only the 1.25 mM treatment for 48 h presented 25% tetraploid plants and in the 8694-15 genotype, the 5.0 mM concentration for 48 h produced 50% tetraploid plants. APM for 24 h enabled the tetraploid plant of the 1304-04 genotype to be obtained, while colchicine for 48 h resulted in tetraploid plants in the 8694-15 genotype.

Further, the efficiency of production of doubled haploid plants in canola (*Brassica napus* L.) breeding programmes is reduced when large numbers of haploid and infertile plants survive until flowering. Cytometry was used to assess ploidy status and predict subsequent fertility of microspore-derived plantlets from three canola genotypes, with or without colchicine treatment of microspore suspensions. Young leaf tissue was sampled from microspore-derived plantlets within 1 week of transfer to soil, and processed immediately by flow cytometry. The process was repeated on the same plants 3–5 weeks later. Of the 519 plants transferred to soil, 57.2% were consistently haploid at both sample times, 33.5% were consistently diploid at both sample times, and the remainder (9.2%) were uncertain or inconsistent in ploidy status across sampling times. Of the 518 plants that survived to flowering, 32.4% were diploid at both times of sampling and fertile (set seed) and 46.3% were haploid at both sampling times and infertile. Another 10.8% were haploid at both sampling times and fertile, but had low pollen viability and seed set, and some were triploid or of uncertain ploidy level. Colchicine treatment of microspore suspensions significantly increased the proportion of diploid plants from 9.7 to 69.7%, with significant variation among genotypes. Evidence from simple sequence repeat marker loci indicated that diploid and fertile plants from the control treatment (no colchicine) were derived from spontaneously doubled haploid gametes, rather than unreduced gametes or somatic tissue. Flow cytometry at the first sample time was very efficient in detecting diploid plants of which 94.2% were subsequently fertile [46].

We conducted a study of the cell cycle of coconut palm tissues cultured in vitro in order to regulate regeneration. Cell nuclei were isolated from various types of coconut palm tissues with and without *in vitro* culture. After the nuclei were stained with propidium iodide, relative fluorescence intensity was estimated by flow cytometry. Characterization of the cell cycle reinforced the hypothesis of a block in the G0/G1 and G1/S phases of the coconut cells. A time-course study carried out on immature leaves revealed that this block takes place gradually, following the introduction of the material *in vitro*. Synchronization of *in vitro*-cultured leaves cells using 60 μM aphidicholin revealed an increase in the number of nuclei in the S phase after 108 h of treatment. The significance of these results is discussed in relation to the ability of coconut tissue cultured in vitro to divide [52].

6.2. Other applications

Cytometry can be used to assess the degree of polysomaty and endoreduplication [48], reproduction pathways [49], and cell cycle [47]. In reference [50] detected mixoploidy (variable amounts of DNA in tissue) and aneuploidy (variations in a small number of chromosomes) by flow cytometry [51].

Several protocols for measuring DNA have been developed, including bivariate analysis related to cytokeratin/DNA analysis/DNA analysis of BrdU and a synthetic nucleoside similar to thymine. These protocols are used to study the cell cycle and to obtain multiparametric measurements of cellular DNA content; they were developed in tandem with commercial software for analyzing the cell cycle [47].

7. Final considerations

Although flow cytometry significantly impacts several fields of plant research, various methodological challenges must be overcome before its potential can be fully realized.

The research group in UFLA's Department of Agriculture consistently attempts to use methodologies for analyzing nuclear DNA content in plants, which removes some technical constraints. We emphasize the importance of research, particularly in disseminating knowledge on best practices, such as standardization type, fluorochrome selection, data presentation, and quality outcome measures.

Author details

Moacir Pasqual, Leila Aparecida Salles Pio,
Ana Catarina Lima Oliveira and Joyce Dória Rodrigues Soares
Federal University of Lavras (UFLA), Department of Agriculture, Lavras, MG, Brazil

8. References

[1] Gucker FT. et al. photoelectric counter for colloidal particles. Am. J. Chem. 69:2422–2431. 1947.

[2] Shapiro HM. The Evolution of Cytometers. Cytometry Part A, v.58A, p.13–20. 2004.

[3] Campos JMS. Obtenção de híbridos hexaplóides e análise genômica de *Pennisetum sp.* por citometria de fluxo. 115p. PhD thesis. Federal University of Lavras. 2007.

[4] Doležel J. Applications of flow cytometry for the study of plant genomes. Journal of Applied Genetics, v.38, p.285–302. 1997.

[5] Galbraith DW et al. Rapid flow cytometric analysis of the cell cycle in intact plant tissues. Science, v.220, p.1049–1051. 1983.

[6] Loureiro JCM et al. Comparison of four nuclear isolation buffers for plant DNA flow cytometry. Annals of Botany 2006b; 98 679–689.

[7] Doležel J. Flowcytometric analysis of nuclear DNA content in higher plants. Hytochemical Analysis 1991; 2(4) 143-154.

[8] Robinson JP. Introduction to flow cytometry. Flow cytometry talks. USA: Purdue University Cytometry Laboratory. http://www.cyto.purdue.edu/flowcyt/ educate/pptslide.html. 2006 (accessed 26 Juny 2011).

[9] Galbraith DW et al. Rapid flow cytometric analysis of the cell cycle in intact plant tissues. Science 1993; 220 1049–1051.

[10] Ochatt SJ. Flow Cytometry in Plant Breeding. Cytometry Part A 2008; 73 581-598.

[11] Loureiro J et al. Two new nuclear isolation buffers for plant DNA flow cytometry: a test with 37 species. Annals of Botany; 2007 100 875–888.

[12] Timbó ALO. Determinação de protocolo para duplicação cromossômica e identificação do nível de ploidia utilizando citometria de fluxo em Brachiaria spp. PhD thesis. Federal University of Lavras; 2010.

[13] Loureiro JCM, Santos C. Aplicação da citometria de fluxo ao estudo do genoma vegetal. Boletim de Biotecnologia 2004; 77 18-29.

[14] Wan Y et al. Ploidy levels of plants regenerated from mixed ploidy maize callus cultures. In Vitro Cellular and Developmental Biology 1992; 28(2) 87-89.

[15] Galbraith DW, Harkins KR, Maddox JM, Ayres NM, Sharma DP, Firoozabady E: Rapid flow cytometric analysis of the cell cycle in intact plant tissues. Science 1983; 4601(220) 1049-1051.

[16] Doležel J, Binarová P, Lucretti S. Analysis of nuclear DNA content in plant cells by flow cytometry. Biologia Plantarum 1989; 31 113–120.

[17] Otto FJ: DAPI staining of fixed cells for high-resolution flow cytometry of nuclear DNA in methods in cell biology. In Methods in cell biology. Edited by Crissman HA, Darzynkiewicz Z . New York: Academic Press,1990;105-110.

[18] Doležel J, Gohde W. Sex determination in dioecious plants Melandrium album and M. rubrum using high-resolution flow cytometry. Cytometry 1995; 19 103–106.

[19] Pfosser, M. et al. Evaluation of sensitivity of flow cytometry in detecting aneuploidy in wheat using disomic and ditelosomic wheat-rye addition lines. Cytometry 1995; 21(4) 387-393.

[20] Marie D, Brown SC. A cytometric exercise in plant DNA histograms, with 2c values for seventy species. Biology of the Cell 1993; 78:41-51.

[21] Doležel J, Bartos J. Plant DNA flow cytometry and estimation of nuclear genome size. Annals of Botany 2005; 95(1) 99-110.

[22] Noirot M. et al. Nucleus– cytosol interactions—A source of stoichiometric error in flow cytometric estimation of nuclear DNA content in plants. Annals of Botany 2000; 86 309–316.

[23] Pinto G et al. Analysis of the genetic stability of *Eucalyptus globulus* Labill. somatic embryos by flow cytometry. Theoretical and Applied Genetics 2004; 109 580–587.

[24] Loureiro JCM et al. Flow cytometric and microscopic analysis of the effect of tannic acid on plant nuclei and estimation of DNA content. Annals of Botany 2006a; 98 515–527.

[25] Walker D, Monino I, Correal E; Genome size in *Bituminaria bituminosa* (L.) C.H. Stirton (Fabaceae) populations: separation of 'true' differences from environmental effects on DNA determination. Environmental and Experimental Botany 2006; 55 258–265.

[26] Doležel J, Greilhuber J. Nuclear Genome Size: Are We Getting Closer? Cytometry Part A 2010; 77 635-642.

[27] Price HJ, Hodnett G, Johnston JS. Sunflower (*Helianthus annuus*) leaves contain compounds that reduce nuclear propidium iodide fluorescence. Annals of Botany 2000; 86 929–934.

[28] Gregory TR. Animal genome size database. http://www.genomesize.com. (accessed 18 December 2009).

[29] Bennett MD, Leitch IJ. Plant DNA C-values database (release 4.0, October 2005). http://data.kew.org/cvalues/. (accessed 02 July 2011).

[30] Doležel J. et al.. Plant genome size estimation by flow cytometry: Inter-laboratory comparison. Annals of Botany 1998; 82 17-26.

[31] Doležel J, Sgorbati S, Lucretti S. Comparison of three DNA fluorochromes for flow cytometric estimation of nuclear DNA content in plants. Physiologia Plantarum 1992; 85 625–631.

[32] Doležel J, Binarová P, Lucretti S. Analysis of nuclear DNA content in plant cells by flow cytometry. Biologia Plantarum 1989; 31(2) 113-120.

[33] Doležel J, Sgorbati S, Lucretti S. Comparison of three DNA fluorochromes for flow cytometric estimation of nuclear DNA content in plants. Physiologia Plantarum, 1992; 85(4) 625-631

[34] Buitendijk JH, Boon EJ, Ramanna MS. Nuclear DNA content in twelve species of *Alstroemeria* L. and some of their hybrids. Annals of Botany 1997; 79 343-353.

[35] Van Dyke WW, Dervan PB. Chromomycin, mithramycin and olivomycin binding sites on heterogeneous deoxyribonucleic acid. Footprinting with (methidiumpropyl-EDTA)iron(II). Biochemistry 1983; 22 2373–2377.

[36] Portugal J, Waring M. Assignment of DNA binding sites for dapi and bisbenzimide (hoeschst 33258). Comparative footprinting study. Biochimea Biophysica Acta 1988; 949 158-168.

[37] Michaelson MJ, Price HJ, Ellison JR, Johnston JS. Comparison of plant DNA contents determined by Feulgen Microspectrophotometry and Laser Flow Cytometry. American Journal of Botany 1991; 78 183-188.

[38] Yanpaisan W, King NJ, Doran PM. Flow cytometry of plant cells with applications in large-scale bioprocessing. Biotechnology Advanced 1999; 17 23–27.

[39] Johnston JS. et al. Reference standards for determination of DNA content of plant nuclei. American Journal of Botany 1999; 86 609–613.

[40] Price HJ, Johnston JS. Influence of light on DNA content of *Helinathus annuus* Linnaeus. Proceedings of the National Academy of Sciences of the USA 1996; 93 11264-11267.

[41] Sliwinska E, Zielinska E, Jedrzejczyk I. Are seeds suitable for flow cytometric estimation of plant genome size? Cytometry Part A 2005; 64 72–79.

[42] Greilhuber J, Temsch E, Loureiro J. Nuclear DNA content measurement. Edited by Doležel J, J Greilhuber J, Suda J, Flow Cytometry with Plant Cells: Analysis of Genes, Chromosomes and Genomes. Weinheim: Wiley-VCH, 2007; 67-101.

[43] Suda J, Kron P, Husband BC, Trávnícek P. Flow cytometry and ploidy: applications in plant systematics, ecology and evolutionary biology. Edited by Doležel J, J Greilhuber J, Suda J, Flow Cytometry with Plant Cells: Analysis of Genes, Chromosomes and Genomes. Weinheim: Wiley-VCH, 2007; 67-101.

[44] Prado MJ, Rodriguez E, Rey L, Gonzalez MV, Santos C, Rey M. Detection of somaclonal variants in somatic embryogenesis regenerated plants of *Vitis vinifera* by flow cytometry and microsatellite markers. Plant Cell Tissue and Organ Culture 2010; 103 49–59.

[45] Rodrigues FA, Soares JDR, Santos RRS, Pasqual M, Silva SOS. Colchicine and amiprophos-methyl (APM) in polyploidy induction in banana plant. African Journal of Biotechnology 2011; 10(62) 13476-13481.

[46] Takahita J, Cousin JA, Nelson MN, Cowling WA. Improvement in efficiency of microspore culture to produce doubled haploid canola (*Brassica napus* L.) by flow cytometry. Plant Cell Tissue and Organ Culture 2011; 104 51–59.

[47] Sandoval A, Hocker V, Verdeil JL. Flow cytometric analysis of the cell cycle in different coconut palm (Cocos nucifera L.) tissues cultured in vitro. Plant Cell Reports 2003; 22 25–31.

[48] Barow M, Meister A. Lack of correlation between AT frequency and genome size in higher plants and the effect of nonrandomness of base sequences on dye binding. Cytometry 2003; 47 1–7.

[49] Matzk F, Meister A, Schubert I. An efficient screen for reproductive pathways using mature seeds of monocot and dicots. The Plant Journal 2000; 21 97–108.

[50] Doležel J. Applications of flow cytometry for the study of plant genomes. Journal of Applied Genetics 1997, 38 285–302.

[51] Roux N. et al. Rapid detection of aneuploidy in Musa using flow cytometry. Plant Cell Reports 2003; 21 483–490.

[52] Nunez R. DNA Measurement and Cell Cycle Analysis by Flow Cytometry. Current Issues in Molecular Biology 2001; 3(3) 67-70.

Chapter 8

Polyamines, Gelling Agents in Tissue Culture, Micropropagation of Medicinal Plants and Bioreactors

Giuseppina Pace Pereira Lima, Renê Arnoux da Silva Campos, Lilia Gomes Willadino, Terezinha J.R. Câmara and Fabio Vianello

Additional information is available at the end of the chapter

1. Introduction

Currently, tissue cultures of species of agricultural importance have wide applicability in industrial production processes. Tissue culture is a name given to a set of techniques that allow the regeneration of cells, tissues and organs of plants, from segments of plant organs or tissues, using nutrient solutions in aseptic and controlled environment. This regeneration is based on the totipotency of plant cells. Totipotency is a capability indicating that plant cells, in different times, may express the potential to form a new multicellular individual. Tissue culture appears to be a good alternative to conventional propagation, requiring less physical space, with high multiplication rate, without incidence of pests and diseases during cultivation, and enabling higher control of the variables involved. Thus, in the *in vitro* environment, with the required stimuli and appropriate conditions, different cell types express different behaviors, possibly leading to cell multiplication and differentiation into a specific tissue, characterized by a form and a function, which may lead to the regeneration of a new individual.

The discovery of this feature in plant cells is indistinguishable from the first studies on tissue culture in the early twentieth century by Heberlandt in 1902, which were followed by the first practical results reported by White in 1934 [1].

Over the years, various tissue culture techniques were developed, being micropropagation, meristem culture and somatic embryogenesis, the most used. The degree of success of any technology employing cultured cells, plant tissues or organs, is mainly dependent on the choice of the nutritional components and growth regulators which control, in a large extent, the developmental *in vitro* pattern. Therefore, it is crucial to evaluate the nutritional and metabolic needs of cells and tissues of each species to be cultivated. In general, the choice of the medium is carried out taking into account, in addition to these needs, the purpose of the *in vitro* cultivation, maximizing plant development.

In general, the culture medium is composed of inorganic salts, reduced nitrogen compounds, a carbon source, vitamins and amino acids. Other compounds may be added for specific purposes, such as plant growth regulators, gelling agents, organic nitrogen compounds, organic acids and plant extracts.

Throughout the history of tissue culture, various kinds of culture media have been developed. However, the MS (Murashige & Skoog) medium [2] is the most widely used for the regeneration of dicots, and therefore it has a great importance in the applications of tissue culture in agriculture.

2. The use of polyamines as growth regulators

Many plant growth regulators have been used in vitro. Generally, literature reports on the use of auxins, cytokinins and gibberellins and different balances of auxins and cytokinins to encourage the development of specific organs. An auxin/cytokinin ratio of 10 induces the rapid growth of undifferentiated callus, a ratio of 100 leads to root development and a ratio of 4 favors development of shoots [3].

However, some studies report on the use of polyamines, as growth regulators. Polyamines (PAs) are low molecular weight aliphatic amines, implicated in various physiological and developmental processes in plants [4], such as growth regulation, cell division and differentiation, and also in the plant response to various sources of stresses. The exogenous application of PAs has been used by many researchers to provide or enhance growth and cell division [4]. The most common types of PAs are spermidine (SPD), spermine (SPM), and their diamine precursor, putrescine (PUT).

In plants, the mechanisms regulating both the biosynthesis and degradation of polyamines are less studied than in other organisms. Some papers proposed the use of exogenous polyamines during the processes leading to *in vitro* plant and callus formation [5,6] and endogenous PAs concentration has been related to several organo-genetic processes[7].

Beneficial effects of polyamines on *in vitro* regeneration and somatic embryogenesis were documented in several crops. Promoter effect of PAs in the conversion of somatic embryos or shoot regeneration has been evidenced in several plant species. The exogenous addition of PAs, mainly SPD or PUT at 2.0 mM concentration, not only elevated the endogenous levels of PAs but also enhanced the frequency of conversion of protocorm-like-bodied (PLBs) to shoots in *Dendrobium huoshanense* [8]. On the other hand, spermidine showed no effect on *Dendrobium* "Sonia". Among various tested polyamines, the maximum number of PLBs was produced with 0.4 mM putrescine treatment. The increase (1.0 mM) or decrease (0.2 mM) of SPD concentration caused a decrease in the production of PLBs. All treatments with spermidine and spermine resulted in the production of less number of PLBs than control. Thus, different responses to the application of different polyamines may occur [9].

High levels of free putrescine were correlated with the ease of cultures to reach stabilization, as observed with juvenile tissues. Adult tissues, containing low levels of putrescine, were difficult to stabilize in culture [10]. Furthermore, it has been reported [11] that the application of putrescine may reduce the production of unwanted ethylene and can enhance morphogenesis.

The exogenous application of polyamines has shown positive effect in the micropropagation of several species, for example, in buds from newly developed shoots, obtained from forced outgrowth of mature field-grown hybrids of hazelnut trees (*Corylus avellana* L.), cultured *in vitro* on MS medium [3] and on a modified Driver and Kuniyuki medium [12] [DKW] DKW/Juglans containing 6.7 μM, 11.1 μM or 15.5 μM N-6-benzyladenine (BA), supplemented with or without a combination of polyamines (0.2 mM putrescine + 0.2 mM spermidine + 0.05 mM spermine). The effects of culture medium and BA were found to be insignificant on explant response. Polyamines were found to have a strong effect on both shoot elongation and on the number of buds per shoot. Polyamines increased both the mean shoot elongation by 83% and the number of buds per shoot by 41%, compared to controls. In the presence of polyamines, shoot elongation continued up to 4.0 cm, while in the absence, the shoot elongation reached only 2.0 cm. Moreover, the results indicate that polyamines added to culture medium can improve the stabilization of cultures and enhance the morphogenic capacity of mature explants [13].

The effects of exogenous polyamines on somatic embryo formation in carrot (*Daucus carota* L.) cells were investigated [14]. The results showed an enhancement of somatic embryo formation following the addition of spermine. However, this effect was not always reproducible. Spm addition increased the DNA content in cells, regardless the development and the lag time of somatic embryo formation, and suppressed the protein secretion from cells.

Many studies have shown the beneficial effects of applying various polyamines on the rooting process. Polyamines have been shown to be key factors, in conjunction with auxins, in the process of adventitious rooting production [15,16]. The *in vitro* rooting of wild-type tobacco (*Nicotiana tabacum* cv. Xanthi) shoots was promoted by polyamines in the absence of any other growth regulator and was inhibited by two inhibitors of polyamine metabolism [17]. According to authors, some arguments causally implicated Put in the inductive rooting phase of poplar: i) the transient increase of Put did not occur in the non-rooting cuttings, ii) it was observed only in the basal rooting zone, iii) inhibitors of Put biosynthesis, such as DFMO and DFMA (α-α-difluoromethylornithine and difluoromethylarginine, respectively) applied prior to, or at the beginning of, the inductive phase, inhibited rooting [18], iv) an inhibitor of Put conversion into Spd and Spm, CHA (cyclohexylamine, an inhibitor of Spd synthase), promoted the accumulation of endogenous Put and favored rooting in the absence of auxin, v) the administration of Put, prior to, or at the beginning of, the inductive phase, stimulated rooting [19]. Some studies correlated this effect with PAs antioxidant properties, such as during the morphogenesis of *Hemerocallis sp.* It was verified that exogenous Put, alone (10 μM Put), with Spd (10 μM Spd + 10 μM Put) or with Spm (10 μM Spm + 10 μM Put), induced the best results, regarding both the number and the height of differentiated shoots. The combination of the three polyamines in culture medium (10 µM SPD + 10 µM SPM + 10 µM PUT) induced the highest percentage of microplant formation, whereas the treatment with SPD and SPM (10 µM SPD + 10 µM SPM) led to the highest amount of necrotic tissue (65%). Spermine alone (10 µM Spm) was effective in controlling oxidative processes. In fact, oxidative stress during *in vitro* cultivation of plants occurs with high frequency [8].

However, in some species, exogenous polyamines do not play a significant role on *in vitro* growth, such as in *Hancornia speciosa*. In this specie, the application of exogenous polyamines did not improve the growth of callus [20]. Conversely, in most studies it has been reported that the use of specific polyamines or polyamine combinations can enhance the performances of tissue culture in some plant species, especially in those with low rooting potential, browning and hyperhydricity.

3. Alternative gelling agents

The *in vitro* cultivation of plant tissues is generally carried out in a solid or semi-solid nutrient medium, using gelling agents. Traditionally, agar is used, which is a polysaccharide extracted from seaweeds. This hydrocolloid is composed of agarobiose (3-β-D-galactopyranosyl-(1,4)-3,6-anhydro-α-L-galactose) [21]. The main differences among different agar-products are due to the impurities, their level and composition, which can vary according to manufacturers. Agar has been widely used since it has convenient gelling properties and stability during tissue culture. In all media used for *in vitro* culture of plants, agar is the major source of unknown variations [22], besides it's a high costs.

Gums, such as gelan, produced by bacteria and commercialized under the name of Gel-Gro®, Gerlite® (Kelko, Merck) and Phyta-gel® (Sigma), are polysaccharides that do not contain contaminating materials. Moreover, these products are used in lesser amount per liter than agar, to obtain the same consistency. They are added to the medium at approximately one fourth the concentration of agar. Furthermore, they appear more transparent. Despite emerging as alternatives to agar, the high cost of these products still limits their use in commercial cultures. These polysaccharides are imported from North America and Europe and therefore this leads to increasing costs for further micropropagation applications.

An alternative for cost reduction is the partial replacement of some of these gelling agents with other polysaccharides. Starch is an inexpensive alternative among studied gelling agents, and its use may reduce the costs of tissue culture. Nevertheless, starch is hydrolyzed by plant amylolytic enzymes during the in vitro culture. To circumvent this occurrence, the increase of air exchange in the bottle and consequently the increase of evaporation of excess water, may reduce this drawback [23].

Tropical countries possess a lot of starchy, little studied, native species, whose characteristics could serve still unfilled market niches. It should be considered that, among five of the raw products used in world for starch production, four are of tropical origin:

potato, cassava, maize and rice. Thus, the possibility to find native starch with specific properties is quite high in tropical regions.

In early experiments, using maize starch as gelling agent, the growth and differentiation of cultured plant cells from tobacco and carrot have been increased. In a medium solidified with starch, cell dry weight increased more than three times with respect to cells grown in a medium gelled with agar [24].

In India, several tissue culture studies were developed using gums and starches derived from tropical species, such as true sago palm (*Metroxylon sagu*) and "isubgol" (*Plantago ovata Forsk.*). Results revealed that these gelling agents were satisfactory in the micropropagation of chrysanthemum (*Dendrathema grandiflora* Tzvelev). Moreover, the cost of cultures was less than 5% of that of commercial agar, resulting in an alternative low-cost gelling agent for industrial scale micropropagation [25].

Isubgol was also successfully used as gelling agent for culture media for *in vitro* germination, formation of aerial parts and roots of *Syzygium cuminii* and cultures of anthers in *Datura innoxia*. Cultures showed similar responses to those observed in medium solidified with agar [26].

Katira gum, derived from the bark of *Cochlospermum religiosum*, was used as gelling agent in the culture medium for micropropagation of *Syzygium cuminii* and for somatic embryogenesis of *Albizzia lebbeck*. Various combinations of agar (0.2-0.6%) and starch (1-3%) were added to increase the firmness of the medium [27].

Guar gum, derived from the endosperm of *Cyamopsis tetragonoloda* and locust bean gum from carob (*Ceratonia siliqua*) (composed of mannose:galactose ratios of 1.6:1 and 3.3:1, respectively) is commercially produced in great amounts. Others, such as tare and fenugreek gums, have also been used [28]. These gums were used as gelling agents for *in vitro* multiplication and regeneration as well for germination of *Linum usitatissimum* and *Brassica juncea*, for the multiplication of aerial parts of *Crateavea nurvala*, and their subsequent rooting, for the *in vitro* androgenesis of anthers on *Nicotiana tabacum*, and for somatic embryogenesis of callus cultures on *Calliandra twedii*. The media were gelled with 2, 3 or 4% gum, as compared to agar (0.9%), and for all species the morphogenetic responses were improved in the medium gelled with guar gum [29].

Xanthan gum has minimal change in viscosity over a wide temperature range, and presents good gelling ability in a broad pH range (3.2-9.8). Xanthan gum, a microbial desiccation-resistant polysaccharide, is commercially produced by aerobic submerged fermentation of *Xanthomonas campestris*. It has a β-(1,4)-D-glucopyranose glucan backbone with side chains of (3,1)-α-linked D-mannopyranose-(2,1)-β-D-glucuronic acid-(4,1)-β-D-mannopyranose, on alternating residues. This gum, cream-colored, odorless, free flowing powder, hydrates rapidly in cold and hot water to give a reliable viscosity, even at low concentrations. It is highly resistant to enzymatic degradation, extremely stable over a wide pH range, and forms highly pseudoplastic aqueous solutions [30].

A gelling agent developed in Brazil (patent PI9003880-0 FAPESP/UNESP) was tested as an alternative to agar in the micropropagation of sweet potato (*Ipomoea batata*). The product consists of a mixture of starch from seeds of pigeonpea (*Cajanus cajan*) and cassava starch (*Manihot esculenta*) at a ratio of 2:3, being used as a gelling agent in the MS culture medium at a concentration of 7%. It was observed that this starch mixture increased the fresh weight of cultures when compared to microplants grown on agar. Thus, this starch mixture represents a good alternative for agar replacement in the micropropagation of sweet potato. Moreover, this substitution reduced by over 94% the final cost of the culture medium, demonstrating a high economic validity [25].

Another successful Brazilian experience deals with xyloglucans, extracted from the seeds of jatoba (*Hymenaea courbaril*) and mixed with agar to prepare a solid medium for the micropropagation of apples. The performance of this new mixture, composed of 0.4% agar and 0.2% xyloglucan (w/v), was verified on *Malus prunifolia Borkh* and cv. *Jonagored* (*Malus domestica*), and compared to the medium solidified with agar (0.6 %, w/v). The growth and the multiplication of shoots were higher in the modified medium. Furthermore, a lower incidence of hyperhydricity and a higher percentage of shoot rooting in the absence of auxin, were observed. When 0.25 mM indole-3-butyric acid (IBA) was added to both the media, the modified medium gave better results in terms of rooting percentage and root quality than the traditional agar medium [31].

The efficacy of the partial substitution of agar by galactomannans (GMs), obtained from seeds of *Cassia fastuosa* (cassia) and *Cyamopsis tetragonolobus* (guargum - a commercial GM), was tested in the micropropagation of strawberry (*Fragaria x ananassa* Duchesne cv. Pelican). GMs were mixed with agar in the proportion of 0.3/0.3 % (w/v) in MS medium, and the performances were compared with the behavior of the medium containing only agar (0.6% w/v). Strawberry shoots, grown in the modified medium, showed both an enhanced cell proliferation and a higher length of roots with respect to controls. These results showed that agar could be partially replaced by GMs, since experimental data confirmed a favorable interaction between the two polysaccharides.

Another study, involving guar gum extracted from the seeds of *Cassia fastuosa* or *Cyamopsis tetragonolobus*, mixed in equal proportions with agar to a final concentration of 0.3% (w/v) for each type of gelling agent, was carried out on the micropropagation of "Durondeau" pear (*Pyrus communis* L. cv. Durondeau). The production of multiple shoots and the formation of roots from shoots were compared with a control media, solidified with agar alone at a concentration of 0.6 % (w/v). In the medium solidified with the mixtures of agar/guar and agar/cassia GM, an increase in the number of regenerated shoots of 32 and 17%, respectively, was obtained. The modified media promoted both a higher number of roots and increased the rooting percentage. A maximum of 91% rooting was obtained in the medium solidified with the agar/cassia GM, containing 9.8 mM indole-3-butyric acid. With this medium, less callus formation at the base of the shoot was also observed [23].

In a study on the micropropagation of african violet, the cultivation of shoots in liquid medium, with cotton wad and different combinations of starch, semolina, potato powder as an alternative to agar, was tested. The highest frequency of regeneration was found in media containing agar (0.8%) or with a combination of starch, semolina, potato powder (2:1:1) and starch (6%) plus agar (0.4%). The maximum numbers shoots was produced in media containing agar (0.8%), the combination of starch (6%), plus agar (0.4%) and in liquid medium with cotton wad substrate. The best shoot proliferation took place in liquid medium with cotton substrate. The results showed that the combination of starch, semolina, potato powder (2:1:1) and 6% starch, plus 0.4% agar, can be suitable alternatives for agar alone in shoot regeneration step, but shoot number will result lower than in agar alone. These options are cheaper than agar [32].

The type of medium (liquid or semi-solid) can directly affect the rooting process. Although the liquid medium positively influences the availability of water, nutrients, hormones and oxygen levels, the vast majority of protocols for micropropagation were established in semi-solid media [33].

Some other agar substitutes have been tested, such as corn starch mixtures (Gelrite®), used for shoot proliferation of apples, pears and red raspberries [34]. However, their use may induce some problems during micropropagation. Gelrite®, for example, causes hyperhydricity and, in some cases, vitrification on regenerated shoots [35]. Hyperhydric shoots are characterized by a translucent aspect due to a chlorophyll deficiency, a not very developed cell wall and high water content. The losses of up to 60% of cultured shoots or plantlets, due to hyperhydricity, have been reported in commercial plant micropropagation [36], reflecting the importance of this problem. The hyperhydricity of shoots was induced in *Prunus avium*, after four weeks multiplication cycles in Gelrite® [37]. This performance has been confirmed by other authors [38]. The vitrification symptoms (thick and translucent stems and leaves, wrinkled and curled leaves) were found on 100 % shoots, after 21 days culture.

Besides the properties of these described substances, it is not always possible to use these agents. In studies with *Eucalyptus grandis* x *E. urophylla* in liquid medium containing aluminum, an acrylic blanket was used to support the shoots, because the addition of $AlCl_3.6H_2O$ decreased medium pH (to around pH 3.5), affecting medium ionic balance, not allowing the gelation of agar [39].

4. Micropropagation of medicinal species

Medicinal and aromatic plants are of great importance for the pharmaceutical industry and traditional medicine in several countries. Just to emphasize the importance of medicinal plants in their various aspects for human health, data from the World Health Organization (WHO) indicated that 80% of the population uses them as basic drugs, in traditional medicine, in the form of plant extracts or their bioactive compounds.

Micropropagation of some medicinal plants has been achieved through the rapid proliferation of shoot-tips and ancillary buds in culture. The success of micropropagation of medicinal species opens perspectives for the production of seedlings, followed by selection of superior genotypes, their clonal multiplication for obtaining genotypic uniformity within the plants to use in plantations at high productivity. Numerous limiting factors have been reported to influence the success of *in vitro* propagation of medicinal plants and, therefore, it is unwise to define a particular protocol for the micropropagation of these species [40].

In fact, a large-scale application depends mainly on the development of an efficient protocol for proliferation, rooting and acclimatization of explants in *ex vitro* conditions.

Jaborandi plant (*Pilocarpus microphyllus*) is a tree, extensively used in the pharmaceutical industry for the manufacture of drugs, used for the treatment of glaucoma. The micropropagation of this specie was conducted on apical explants grown in MS medium, supplemented with 6.66 mM benzyl-aminopurine (BAP) [41]. *Echinodorus scaber* is an herbaceous water plant with cleansing, anti-ophidic, diuretic, anti-rheumatic and anti-inflammatory effects, used in traditional medicine to normalize uric acid, treating gout and osteoarthritis, and also used for the manufacturing of soft drinks. The protocol for this specie consists in the inoculation of nodal explants in MS medium, supplemented with 1 mgL^{-1} BAP [42].

For the micropropagation of *Tournefortia paniculata*, a shrub whose leaves are used as decoction for diuretic and urinary infections, the shoots were grown on WPM (Lloyd & McCown) medium, supplemented with 1 mg L^{-1} BAP, and shoot rooting was induced in the same medium, without growth regulators. For acclimatization, microplants were placed in a commercial substrate, and cultured in a greenhouse, and then transplanted in the field [43].

Lipia gracilis plant is rich in essential oils, containing thymol and canvacrol, as main chemical compounds, responsible for the proven antimicrobial activity. A protocol for its micropropagation has been defined, using MS medium supplemented with 1 mg L^{-1} BAP [44].

The proliferation success, expressed as number of new produced shoots, is the most important parameter that should be optimized *in vitro*, since often the proliferation process is slow and, as consequence, the number of produced plants is limited. This feature can also be combined to obtain secondary metabolites. Micropropagation ensures a uniform production, which cannot be guaranteed by cultivation of seeds of allogamous plants, due to the high genetic variability of offsprings, which cannot guarantees the uniformity on types and levels of metabolites. Several protocols for medicinal plants have been described [42], but many species still require further studies, in particular, to increase or maintain the production of desirable metabolites.

The micropropagation of medicinal plants is particularly important if the plant produces few seeds, seeds with low germination potential and needs special care to reach the mature stages. As an example, *Jatropha elliptica*, an herbaceous native species in Brazilian cerrado with purgative properties, used in the treatment of severe itching, syphilis and treatment of snake bites, presents low seed germination potential, about only 6%. The developed micropropagation protocol consists of MS medium, supplemented with 0.5 µmol L^{-1} indolacetic acid (IAA) and 1 µmol L^{-1} BAP for the multiplication step, and 5 µmol L^{-1} naphthalene acetic acid (NAA) for shoot rooting [45].

An efficient micropropagation protocol has been proposed for *Stevia rebaudiana*, an important antidiabetic medicinal plant. The highest frequency (94.5%) of multiple shoot regeneration and maximum number of shoots (15.7 shoots per explant) were obtained on MS medium, supplemented with 1.0 mg L^{-1} BAP. Furthermore, the *in vitro* derived nodal explants, grown in the same nutritional conditions, produced a total of 123 shoots per explant after only three subcultures. The highest frequency of rooting (96%) was obtained on half-strength MS medium, in the presence of 0.4 mg L^{-1} NAA. The rooted plantlets were successfully transferred into plastic cups, containing sand and soil, in the ratio of 1:2, and subsequently transferred in the greenhouse [46].

Some medicinal plants are "recalcitrant" to vegetative propagation, and in these cases, the technique of *in vitro* cultivation, and the use of different plant growth regulators, can synergistically help plant production. *Byrsonima intermedia* A. Juss is a shrub found in Brazilian cerrado, whose husk presents medicinal activities in diarrheas and dysenteries. The *Byrsonima* genus presents low germination rate and slow plantlet emergency, which makes sexual propagation difficult. The effect of different concentrations of 2,4-dichlorophenoxyacetic acid (2,4-D), thidiazuron (TDZ) and BAP in callus formation, was evaluated. The results demonstrated that there was no formation of callus in leaf explants maintained in the absence of 2,4-D, and that the addition of TDZ and BAP had no influence in the callogenetic process. For callus induction and proliferation, the results suggested the use of MS medium, supplemented with 1.0 mgL^{-1} 2,4-D, maintaining the explants in the dark [47].

Another species, presenting difficulties in spreading, is the Brazilian ginseng (*Pfaffia tuberosa*), whose roots are used to produce drugs and dietary supplements. A protocol for its micropropagation was developed using nodal explants cultivation in MS medium supplemented with 1 mmolL^{-1} thidiazuron (TDZ), followed by shoot subculture on the same medium, lacking of growth regulators. The methodology was proven to be valid, due to the high rate of multiplication, good development of shoots, roots and great adaptation to *ex vitro* conditions [48].

The use of tissue culture techniques is also of great importance in the development of new plant products, by preserving germplasms. Micropropagation is a useful technique to preserve the gene pool, especially in cases where small amounts of viable seeds are produced. As an example, canopy (*Canopy macrospyphonia*), a medicinal species, considered highly vulnerable and widely used by traditional communities inhabiting Brazilian cerrado.

For this species, micropropagation would be an alternative to avoid genetic erosion. Thus, a protocol for its *in vitro* micropropagation and conservation in germplasm bank was developed. It was found that half diluted MS medium (MS/2), without growth regulators, promoted the proliferation of shoots (4 per bud), shoots elongation (5.2 cm), number of buds (6-8) and reduced vitrification (4%). Seedlings were grown for three months in MS/2 culture medium, supplemented with 2% sucrose, 4% mannitol, 2 mgL^{-1} calcium pantothenate, under germplasm Bank conditions, showing 40% survival. Regarding substrates, seedlings grew better in Plantmax®, where 40% of the seedlings survived and the majority showed root formation [49].

Other plants threatened with extinction can be grown *in vitro*, to maintain a germplasm bank, and to guarantee the possibility of increasing their population, as is the case of *Rhinacanthus nasutus* (L.) Kurz., used as a potent ethno-medicinal plant for various diseases, including cancer. Since the plant is collected throughout the year for roots and leaves, it increasingly disappeared from its natural habitat, and, as a consequence, its distribution in nature is reduced to an alarming rate. Natural resurgence of this plant is through seeds and stem cuttings. Low seed germination and viability is another factor that hampers the natural propagation of this plant. The possibility to regenerate whole plants from leaf derived callus was reported. High numbers of shoots (36.5) were obtained through an intervening callus phase. The composition of the media had a significant effect on both percentage of callus formation and subsequent plantlet regeneration. Optimum callus induction (98.8%) was obtained in MS medium, supplemented with kinetin (Kn) (4 mgL^{-1}) and indolbutiric acid (IBA) (0.5 mgL^{-1}). However, the highest callus regeneration (96%) was observed in MS medium, supplemented with 3 mgL^{-1} Kn and 0.5 mgL^{-1} NAA. An important finding in this study was that the ISSR (Inter Simple Sequence Repeat) analysis of *R. nasutus* results in shoots with low degree of variation and statistically not significant differences [50].

Cytotoxic activities of plant extracts against human tumor cell lines can be studied utilizing micropropagation. The stabilization of shoots and roots of *Cistus creticus* subsp. *creticus* L. in *in vitro* culture in solid or liquid MS medium, without addition of plant growth regulators, was studied [51]. Authors verified the presence of labdane diterpenes in shoot extracts, but they were absent in root extracts. The cytotoxic activity of shoot extracts, by sulforhodamine B (SRB) assay on five human cancer cell lines, was verified. In particular, active extracts showed cytotoxic activity on HeLa (cervix), MDA-MB-453 (breast) and FemX (melanoma) cancer cells, with IC$_{50}$ reaching 80.83 mg mL^{-1} on HeLa, 76.18 mg mL^{-1} on MDA-MB-453, and 87.52 mg mL^{-1} on FemX cells, respectively.

A wide application of micropropagation techniques on medicinal plants for clonal propagation and production of virus free plants, if characterized by low immunity to infections, was reported. This is the case of *Aloe vera*, a species traditionally propagated by seedling production, where there is a high probability of disease occurrence in planting material, depending on the injury done to mother plants, seedlings and lateral shoots, at the time of harvest. Thus, micropropagation presents the advantage of a system with high genetic quality and health, and high income [52].

5. Biochemical responses of plants cultivated in vitro

The *in vitro* culture can be used to study the biochemical mechanisms of plant survival, as a function of medium changes, which may produce different biochemical and physiological responses. The *in vitro* cultivation of plants can induce different responses according to the medium used, such as the presence of regulators, gelling agents, minerals, and other components. The concentration of many plant metabolites responds to these changes more or less quickly, but all endogenous molecules may show variations, mainly due to some kind of stress. Stress intensity (pressure to change exerted by a stressor) is not easily quantified. Stress could occur at a low level, creating conditions that are marginally non-optimal, with little expected effect. However, if this mild stress continues for long time, becoming a chronic stress, the physiology of plants is likely to be altered [53].

Tissue culture has been used as a model to study biochemical responses to different stress types on medicinal plants, as reported for oregano (*Origanum vulgare L.*) by using shoot buds as potential model system for studying carbon skeleton diversion from growth to secondary metabolism, as adaptive response to nutrient deficiency. Nutritional stress caused a moderate increase of constitutive free proline, and exogenous proline affected growth and phenolic antioxidant content of oregano shoots, compared to control. The role of proline, and its association to redox cycles, can be considered as a form of metabolic signaling, based on the transfer of redox potential amongst interacting cell pathways, which in turn elicits phenolic metabolism, stimulating carbon flux through pentose phosphate oxidative pathway [54]. Other studies also showed changes in response to mineral stresses, as the study carried out using micropropagation on *Eucalyptus grandis x E. urophylla*, grown with the addition of aluminum (6.75, 13.5 and 27 mgL^{-1} of AlCl$_3$.6H$_2$O). It was observed that the metal affected the ionic equilibrium of the culture medium, the morphology of the shoots, and that the reduced medium pH induced an increase in polyamines content and a higher acid phosphatase activity [41].

As mentioned above, hyperhydricity is another problem that can occur using Gelrite®. This was observed in *P. avium* shoots, causing a production of higher amounts of ethylene, polyamines, and proline, which are substances considered as stress markers. The higher activity of glutathione peroxidase (GPX, EC 1.11.1.9), involved in organic hydroperoxide elimination, suggested an increased production of these last compounds in hyperhydric state [39].

Plant cell culture is a methodology, which can be used to study or to produce some active metabolites, such as polyphenols. *In vitro* culture can be used to explore new industrial, pharmaceutical and medical potentialities, such as the production of secondary metabolites, like flavonoids. This technique was applied to the investigation of *Coriandrum*, and a detailed analysis of individual polyphenols on *in vivo* and *in vitro* grown samples, was performed. *In vitro* samples also gave a high diversity of polyphenols, being C-glycosylated apigenin (2983 mg kg^{-1} d.w.) the main compound. Anthocyanins were found only in clone

A, certainly related to purple pigmentation, and peonidin-3-O-feruloylglucoside-5-O-glucoside was the major anthocyanin found (1.70 mg kg^{-1} d.w.) [55].

These studies show the potential of tissue cultures in plant tissue technology aimed to the production of antioxidants compounds. Therefore, even with the increase of these compounds, plants may present reduced growth, or other problems, such as hyperhydricity. On the other hand, studies involving bioreactors can be an alternative for the production of secondary metabolites.

6. Bioreactors

The use of bioreactors in laboratories and biofactories is already a reality and the trend of its increasing application is indisputable. While conventional micropropagation uses small flasks, with a small number of plants per flask and requires intense manipulation of the cultures, then, involving a large amount of skilled work, the bioreactor uses large bottles containing liquid medium with large amounts of plants, which reduces significantly the demand of skilled operators. Bottles used in conventional micropropagation typically contain less than 0.5 L of culture medium, while bioreactors, on the other hand, may contain amounts ranging from 1.5 to 20 L [56].

Pioneering studies, published in the 90's, showed the superiority of bioreactors on plant multiplication rates, when compared to conventional systems with semi-solid or liquid medium. In the propagation of pineapple (*Ananas comosus*), the multiplication rate in bioreactors was four times higher than that obtained in conventional systems [57]. In banana (*Musa acuminata*) this multiplication rate was five times higher [58] and in sugarcane (*Saccharum edule* Hassk.), it was six [59].

Bioreactors are used in the micropropagation of several crops, including ornamental and medicinal plants, vegetables and fruits. Studies showed that more than 40 plant species are commercially propagated in bioreactors [60]. By cultivation in bioreactors, different plant parts can be obtained, such as buds, somatic embryos, bulbs, shoots, calluses, protocorm and others. Bioreactors are currently being used for commercial micropropagation in U.S., Japan, Taiwan, Korea, Cuba, Costa Rica, Netherlands, Spain, Belgium, France [61], and Brazil (http: biofrabricasdemudas.blogspot.com).

The design of the first bioreactors, used for propagation of plants, was derived from fermenters used for cultivation of bacteria and fungi cultivation for industrial purposes. Currently, there are several types of bioreactors developed specifically for *in vitro* plant cultivation [62]. Among the different types of plant micropropagation bioreactors, the Temporary Immersion Bioreactor (TIB), developed in the mid 90's, stands out [61]. This type of bioreactor consists of two glass or plastic vessels, one of which contains the plant material in culture (plant culture vessel), while the other stores the culture medium. The transfer of the medium to the flask containing the plant material occurs through silicone hoses, driven by positive pressure of an air pump. Generally, explants are immersed in the culture medium temporarily. The soaking process occurs at regular intervals of predetermined time. When the time of immersion is finished, the air pump starts and a solenoid valve drive the air flow to the opposite direction. The circulating air in the system is sterilized by passing through filters with pores of 0.22 to 0.44 μm, coupled to silicone hoses (Figure 1).

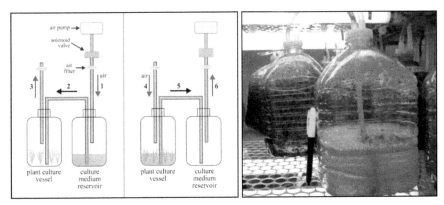

Figure 1. Temporary immersion bioreactor (TIB): The transfer of the medium to the plant culture vessel occurs by a positive pressure of the air pump and the explants are temporarily immersed, after this, the air pump starts and the medium returns to the culture medium reservoir (A) scheme, (B) picture.

The main advantages of temporary immersion bioreactors in plant propagation process include: (1) the liquid medium is in contact with the entire surface of the explants (leaves, roots; etc.), increasing the absorption surface of nutrients and growth regulators; (2) the forced aeration provides excellent oxygen supply and prevents the buildup of harmful gases, resulting in a better crop growth; (3) the movement of explants inside the bioreactor results in reduced apical dominance expression, favoring the proliferation of axillary buds; (4) significant reduction in manpower due to the lower handling of vessels, labeling, etc.; (5) the production of a large number of seedlings, favoring large scale production [59, 63, 64]. The high rates of multiplication, the rapid growth of crops, and the reduction of the need for labor and cost for the culture medium, by avoiding gelling agents, make the technology of micropropagation in temporary immersion bioreactors a conventional static medium in liquid or semi-solid systems.

Author details

Giuseppina Pace Pereira Lima
Institute of Biosciences, São Paulo State University, Botucatu, São Paulo, Brazil

Renê Arnoux da Silva Campos
Mato Grosso State University, Cáceres, Mato Grosso, Brazil

Lilia Gomes Willadino
Federal Rural University of Pernambuco, Recife, Pernambuco, Brazil

Terezinha J.R. Câmara
Federal Rural University of Pernambuco, Recife, Pernambuco, Brazil

Fabio Vianello
University of Padua, Padova, Italia

7. References

[1] Bhojwani SS, Razdan MK. Plant tissue culture: Theory and Pratice, a Revised Edition. Amsterdan, Elsevier Science Publishers; 1996, 767p.

[2] Murashige T, Skoog F. A revised medium for rapid growth and bioassays with tobacco tissue cultures. Physiologia Plantarum 1962;15: 473-497.

[3] Murashige T. Plant growth substances in commercial uses of tissue culture. In: Skoog F, (ed.) Plant Growth substances. Berlin:Springer-Verlag; 1980. p.426-34.

[4] Viu AFM, Viu MAO, Tavares AR, Vianello F, Lima GPP. Endogenous and exogenous polyamines in the organogenesis in Curcuma longa L. Scientia Horticulturae 2009; 121: 501-504.

[5] Takeda T, Hayakawa F, Oe K, Matsuoka H. Effects of exogenous polyamines on embryogenic carrot cells. Biochemical Engineering Journal 2002;12: 21–28.

[6] Debiasi C, Fraguas CB, Lima GPP. Study of polyamines in the morphogenesis *in vitro* of *Hemerocallis* sp.Ciencia Rural 2007; 37(4): 1014-1020.

[7] Francisco AA, Tavares AR, Kanashiro S, Ramos PRR, Lima GPP. Plant growth regulators in polyamines endogenous levels during the development of taro cultivated in vitro. Ciencia Rural 2008; 38(5): 1251-1257.

[8] Wang Y, Luo J-P, Wu H-Q, Jin H. Conversion of protocorm-like bodies of *Dendrobium huoshanense* to shoots: The role of polyamines in relation to the ratio of total cytokinins and indole-3-acetic acidindole-3-acetic acid. Journal of Plant Physiology 2009;166: 2013-2022.

[9] Saiprasad GVS, Raghuveer P, Khetarpal S, Chandra R. Effect of various polyamines on production of protocorm-like bodies in orchid - *Dendrobium* 'Sonia'. Scientia Horticulturae 2004;100: 161–168.

[10] Rey M, Tiburcio, AF, Díaz-Sala C, Rodírguez R. Endogenous polyamine concentrations in juvenile, adult and *in vitro* reinvigorated hazel. Tree Physiology 1994; 14(2): 191-200.

[11] Bais HP, Ravishankar GA. Role of polyamines in the ontogeny of plants and their biotechnological applications. Plant Cell, Tissue and Organ Culture 2002; 69: 1–34, 2002.

[12] Driver JA, Kiniyki AH. In vitro propagation of Paradox walnut rootstock. HortiScience 1984; 19: 507-509.

[13] Nas MN. Inclusion of polyamines in the medium improves shoot elongation in hazelnut (*Corylus avellana* L.) [micropropagation. Turkish Journal of Agriculture and Forestry 2004;28: 189-194.

[14] Takeda T, Hayakawa F, Oe K, Matsuoka H. Effects of exogenous polyamines on embryogenic carrot cells. Biochemical Engineering Journal 2002;12: 21–28.

[15] Gaspar Th, Kevers C, Hausman JF. Indissociable chief factors in the inductive phase of adventitious rooting. In: Altman A, Waisel Y. (eds.) Biology of Root Formation and Development, New York: Plenum Press; 1997, p.55–63.

[16] Kevers C, Bringaud C, Hausman JF, Gaspar Th. Putrescine involvement in the inductive phase of walnut shoots rooting in vitro. Saussurea 1997;28: 50–57.

[17] Faivre-Rampant O, Kevers C, Dommes J, Gaspar T. The recalcitrance to rooting of the micropropagated shoots of the*rac* tobacco mutant: Implications of polyamines and of the polyamine metabolism. Plant Physiology and Biochemistry 2000; 38(6): 441-448.

[18] Hausman JF, Kevers C, Gaspar Th. Auxin-polyamine interaction in the control of the rooting inductive phase of poplar shoots in vitro. Plant Science 1995;110: 63–71.

[19] Hausman JF, Kevers C, Gaspar Th. Involvement of putrescine in the inductive rooting phase of poplar shoots raised in vitro. Physiologia Plantarum 1994;92: 201–206.

[20] Fraguas CB, Villa F, Lima GPP. Evaluation of exogenous application of polyamines on callus growth of mangaba tree (*Hancornia speciosa* Gomes). Revista Brasileira de Fruticultura 2009; 31(4): 1206-1210.

[21] Lucyszyn N, Quoirin M, Koehler HS, Reicher F, Sierakowski M-R. Agar/galactomannan blends for strawberry (*Fragaria x ananassa* Duchesne) cv. Pelican micropropagation. Scientia Horticulturae 2006;107: 358–364.

[22] Scholten HJ, Pierik RLM. Agar as a gelling agent: chemical and physical analysis. Plant and Cell Reports. 1998.17: 230–235.

[23] Pinho, RS. Comparison of agar and starch as gelling agents in the micropropagation of sweet potato (*Ipomea batatas* (L.) Lam). Master Thesis. Universidade Estadual Paulista, 2002.

[24] Henderson WE, Kinnersley AM. Corn starch as an alternative gelling agent for plant tissue culture. Plant Cell, Tissue and Organ Culture 1988;15 (1) 15-22.

[25] Bhattacharya P, Satyahari D, Bhattacharya BC. Use of low-coast gelling agents and support matrices for industrial scale plant tissue culture. Plant Cell, Tissue and Organ Culture 1994; 37 (1): 15-23.

[26] Babbar SB, Jain N. 'Isubgol" as an alternative gelling agent in plant tissue culture media. Plant Cell Reports 1998; 17: 318-322.

[27] Jain N, Babbar SB. Gum katira – a cheap gelling agent for plant tissue culture media. Plant Cell, Tissue and Organ Culture 2002; 71(1): 223-229.

[28] Maier H, Anderson M, Kar C, Maqnuson K, Whistler RL. Guar, locust bean, tara and fenugreek gums. In: Whistler, R.L., Bemiller, J.N.(eds.) Industrial Gums: Polysaccharides and their Derivatives. 3rd ed. New York:Academic Press; 1993 p215-218.

[29] Babbar SB, Jain N, Walia N. Guar gum as a gelling agent for plant tissue culture media. In vitro Cellular & Developmental Biology-Plant 2005;41: 2005.

[30] Babbar SB, Jain N. Xanthan gum: an economical partial substitute for agar in microbial cultue media. Current Microbiology 2006;52: 287–292.

[31] Lima-Nishimura N, Quoirin M, Naddaf YG, Wilhelm M, Ribas LLF, Sierakowski MR. A xiloglucan from seeds of the native Brazilian Species. Plant Cell Reports 2003;21 (5): 402-407.

[32] Sharifi1 A, Moshtaghi N, Bagheri A. Agar alternatives for micropropagation of African violet (*Saintpaulia ionantha*). African Journal of Biotechnology 2010;9 (54): 9199-9203.

[33] Souza AV, Pereira AMS. In vitro cultivated plant's rooting. Brazilian Journal of Medicinal Plants 2007;9 (4): 103-117.

[34] Zimmerman RH, Bhardwaj SV, Fordham IM. Use of starch-gelled medium for tissue of some fruit crops. Plant Cell Tissue and Organ Culture 1995;43: 207–213.

[35] Pasqualetto PL, Zimmerman RH, Fordham I. The influence of cation and gelling agent concentration on vitrification of apple cultivars in vitro. Plant Cell Tissue and Organ Culture 1988;14: 31–40.

[36] Pâques M. Vitrification and micropropagation: causes, remedies and prospects. Acta Horticulturae 1991;289: 283-290.

[37] Franck T, Crèvecoeur M, Wuest J, Greppin H, Gaspar T. Cytological comparison of leaves and stems of Prunus avium L. shoots m with agar or gelrite, Biotechnic and Histochemistry 1998;73: 32–43.

[38] Franck T, Kevers C, Gaspar T, Dommes J, Deby C, Greimers R, Serteyn D, Deby-Dupont G. Hyperhydricity of *Prunus avium* shoots cultured on gelrite: a controlled stress response. Plant Physiology and Biochemistry 2004;42: 519–527.

[39] Basso LHM, Lima GPP, Gonçalves AN, Vilhena SMC, Padilha CCF. Effect of aluminium on the free polyamines content and acid phosphatase activity during the growth of *Eucalyptus grandis* x *E. urophylla* shoots cultivated in vitro. Scientia Forestalis 2007; 75: 9-18.

[40] Rout GR, Samantaray S, Das P. In vitro manipulation and propagation of medicinal plants Biotechnology Advances 2000;18: 91–120.

[41] Sabá RT, Lameira AO, Luz JMQ, Gomes APR, Innecco R. Micropropagation of the jaborandi. Horticultura Brasileira 2002;20 (1): 106-109.

[42] Pereira FD, Pinto JE, Cardoso, MG, Lameira OL. Propagation in vitro of "chapéu-de-couro" (*Echinodoruscf. scaber* Rataj), a medicinal plant.Ciência e Agrotecnologia 2000;24 (1) 74-80.

[43] Bertolucci SK, Pinto JE, Cardoso MG. et al. Micropropagation of *Tournefortia* cf *paniculata* Cham. Brazilian Journal of Medicinal Plants 2000; 3 (1): 43-49.

[44] Marinho MJM, Albuquerque CC, Morais MB, Souza MCG, Silva KMB. Establishment of protocol for *Lippia gracilis* Schauer micropropagation. Brazilian Journal of Medicinal Plants 2011;13 (2): 246-252.

[45] Campos RA, Añez LM, Dombroski JL, Dignart SL. Micropropagation of *Jatropha elliptica* (Pohl) Müll. Arg. Brazilian Journal of Medicinal Plants 2007; 9 (3): 30-36.

[46] Thiyagarajan M, Venkatachala P. Large scale *in vitro* propagation of *Steviarebaudiana* (bert) for commercial application: Pharmaceutically important and antidiabetic medicinal herb. Industrial Crops and Products 2012; 37(1): 111-117.

[47] Nogueira RC, Paiva R, Oliveira LM, Soares GA, Soares FP, Castro AHF, Paiva PDO. Calli induction from leaf explants of murici-pequeno (*Byrsonima intermedia* A. Juss.). Ciência & Agrotecnologia de Lavras 2007; 31(2): 366-370.

[48] Flores R, Nicoloso FT, Maldaner J. Rapid clonal micropropagation of *Pfaffia tuberosa* (Spreng.) Hicken. Brazilian Journal of Medicinal Plants 2007; 9 (1): 1-7.

[49] Martins LM, Pereira MAS, França SC, Berton BW. Micropropagation and conservation of *Macrosyphonia velame* (St. Hil.) Muell. Arg. *in vitro* germoplasm bank. Ciência Rural 2011; 41 (3): 454-458.

[50] Cheruvathur MK, Sivu AR, Pradeep NS, Thomas TD. Shoot organogenesis from leaf callus and ISSR assessment for their identification of clonal fidelity in *Rhinacanthus nasutus* (L.) Kurz., a potent anticancerous ethnomedicinal plant Industrial Crops and Products 2012; 40: 122–128.

[51] Skoric M, Todorovic S, Gligorijevic N, Jankovic S, Ristic M, Radulovic S. Cytotoxic activity of ethanol extracts of *in vitro* grown *Cistuscreticus* subsp. *creticus* L. on human cancer cell lines. Industrial Cropos and Products 2012; 38: 153-159.

[52] Araujo P S, Silva JM, Neckel CA. Micropropagação de babosa (*Aloe vera* – Liliaceae). Biotecnologia 2002;25: 54-57.

[53] Gaspar T, Franck T, Bisbis B, Kevers C, Jouve L, hausman JF, Dommes J. Concepts in plant stress physiology. Application to plant tissue cultures. Plant Growth Regulation 2002; 37(3): 263-285.

[54] Lattanzio V, Cardinali A, Ruta C, Fortunato IM, Lattanzio VMT, Linsalata V, Cicco N. Relationship of secondary metabolism to growth in oregano (*Origanum vulgare* L.) shoot cultures under nutritional stress. Environmental and Experimental Botany 2009; 65(1): 54-62.

[55] Barros L, Dueñas M, Dias MI, Sousa MJ, Santos-Buelga C, Ferreira ICFR. Phenolic profiles of in vivo and in vitro grown *Coriandrum sativum* L. Food Chemistry (2012);132: 841–848.

[56] Takayama, S. & Akita, M. The types ofbioreactors used for shoots and embryos. Plant Cell, Tissue and Org.Cult. 39:147-156, 1994.

[57] Escalona, M.; Lorenzo, J.C.; Gonzalez, B.L.; Danquita, M.; Gonzales, J.L.; Desjardins, Y.; Borroto, C.G. Pineapple (Ananas comosus (L.) Merr.)micropropagation in temporary immersion systems. Plant CellReports, v.18, p.743-748, 1999.

[58] Alvard, D.; Cote, F.; Teisson, C. Comparasion of methodsof liquid medium culture for banana micropropagation. PlantCell, Tissue and Organ Culture, Dordrecht, v. 32, p. 55-60, 1993.

[59] Lorenzo, J. C.; González, B. L.; Escalona, M.; Teisson, C.; Espinosa, P.; Borroto, C. Sugarcane shoot formation in an improved temporary immersion system. Plant Cell, Tissue and Organ Culture, Dordrecht, v. 54, p.197-200, 1998.

[60] Mehrotra, S.; Goel, M.K.;Kukreja, A.K.; Mishra, B.N. Efficiency of liquid culture systems over conventional micropropagation: A progress towards commercialization. African Journal of Biotechnology, v. 6, p. 1484-1492, 2007.

[61] Ziv, M. Bioreactor technology for plantmicropropagation. Horticultural Reviews, v. 24, p. 1-30, 2010.

[62] Yesil-Celiktas, O.; Gurel, A; Vardar-Sukan, F. Large scale cultivation of plant cell and tissue culture in bioreactors. Kerala, Transworld Research Network, 2010. 54p.

Chapter 9

Tissue Culture Techniques for Native Amazonian Fruit Trees

Moacir Pasqual, Edvan Alves Chagas,
Joyce Dória Rodrigues Soares and Filipe Almendagna Rodrigues

Additional information is available at the end of the chapter

1. Introduction

The fruits of the Amazon have attracted great interest in recent years, both nationally and internationally, according to its exotic flavors and pleasant and varied ways to use its pulp by agribusiness [1], pharmaceutical industry [2], high vitamin and antioxidant content [3].

In recent decades, the production of native fruits of the Amazon showed significant growth, mainly due to expansion of area for fruit production. It is noteworthy that this activity has had little impact on native vegetation, since most of the orchards were planted in areas previously occupied by other crops for market problems or environmental issues and pressure for sustainable agriculture, ceased to be interesting for farmers [4].

The Amazon forest has large number of non-domesticated fruit species and a minority being exploited through crop in place of natural occurrence [5]. According to the Brazilian Yearbook of Fruit [6], explored the country are 500 varieties of edible fruit-producing species native and exotic, and of these, 220 are still as untamed. The high rate of destruction of biomes, together with the predatory extraction, result in loss of genetic material of desirable characteristics, [7], with potential for use in food, as an ornamental or in pharmaceutical production can never be known. It is therefore essential to know these species and their growing needs for exploitation on a commercial scale, a rational and sustainable.

However, little efficient production technology and knowledge of native Amazonian fruit tree species exist. Low orchard productivity indicates that Amazonian fruit is underused; underuse, in turn, has hindered its cultivation. In [8] notes that many of the current fruit production systems were developed empirically, which required technology that ensured greater productivity, sustainability, and profitability than the older production systems.

Such technology involves crop management, production of reproducible seedlings, and distribution of the seedlings to farmers. Therefore, the first step in domesticating native fruit species and in introducing them to commercial cultivation is developing seedling propagation techniques.

The sexual propagation of native plants for use in horticulture, is not advantageous because it results in populations with wide variation in the period of maturation. In addition, some species have seeds with dormancy, which compromises the germination and seedling production on a commercial scale, or the recalcitrance of the seeds, preventing their storage for extended periods.

The method of propagation of fruit species most commonly used grafting. This technique, like other methods of vegetative propagation allow the cloning of selected plants directly from nature or from artificial hybridizations, maintaining their desirable traits. Vegetative propagation results in high quality seedlings in orchards and more uniform and earlier, with higher productivity and better quality of fruits [9]. Besides these advantages, the graft is also advantageous because it allows greater control of plant height facilitating the management and harvesting, and the formation of plants with resistance to pathogens in soils and drought tolerant.

2. Tissue culture techniques for native Amazonian fruit trees

For decades numerous studies have been developed with the objective of establishing protocols for micropropagation of fruit species for clonal multiplication of superior individuals, in view of the numerous advantages it offers in vitro propagation.

Besides the application of tissue culture techniques for plant propagation of high agronomic value carriers or rare genes and those at risk of extinction, the technique is also used for cleaning clonal plants through tissue culture. In plant breeding programs, the tissue culture can be used to reduce the time for development of new cultivars and expansion of genetic variability. Among the tissue culture techniques most widely used in plant breeding, there may be mentioned: in vitro selection of resistant /tolerant to various stress factors of the genetic variability from pre-existing or induced somaclonal variation or by use of mutagenic, haploidization, somatic hybridization, and rescue of zygotic embryos obtained from crosses between different species or genera of plants [10]. Tissue culture also stands as an adjuvant in introgression of genes of agronomic interest through genetic engineering.

An application of tissue culture which is very important, especially for native plants is the maintenance and storage of germplasm. However, in many other species, tissue culture techniques have been also widely used for studying the metabolism, physiology, development and reproduction of plants with desirable commercial property of interest [11], like the nutraceutical and pharmaceutical products.

Studies on the application of tissue culture techniques in fruit tree species native to the Amazon, are still incipient. Except for a few species in which native tissue culture studies are in an advanced stage, such as *Theobroma cacao* and *T. grandiflorum*, for most species have been developed with the aim of improving methods for initial establishment of cultures in vitro (Table 1).

Species	Purpose of the study	Author
Açai (*Euterpe oleracea.*)	Culture of embryos	[12]
Araça (*Psidium spp.*)	Establishment of culture	[13]
Bacuri (*Platonia insignis*)	Establishment of culture	[14], [15]
Cacau (*Theobroma cacao*)	somatic embryogenesis	[16], [17]
	Multiplication of shoots	[18]
	In vitro establishment	[19], [14]
Caja (*Spondias mombin*)	Culture of embryos and callus	[20]
Camu-camu (*Myrciaria dubia*)	Effect of culture media on morphogenetic responses	[21], [22], [23]
Cupuaçu (*Theobroma grandiflorum*)	In vitro establishment	[24], [25]
	Callus	[26], [27]
	somatic embryogenesis	[28], [29]
Inga (*Inga vera*)	Establishment of culture	[30]
Murici (*Byrsonima basiloba*)	direct organogenesis	[7]
Murmuru	In vitro propagation and somatic embryogenesis	[9]
Pupunha (*Bactris gasipaes*)	Cultivation of embryos	[31]

Table 1.

3. *In vitro* establishment and micropropagation

Micropropagation is a high-impact plant tissue culture technique that is more consistent than other tissue culture methods. Micropropagation involves a high reproduction rate over a short time period, which produces plants with excellent phytosanitary quality. The technique involves various steps, from aseptic *in vitro* culture establishment to rooting, culminating in seedling acclimatization [32].

Most studies on the tissue culture of native Amazonian plant species have sought to improve methodologies for initial culture establishment. Disinfection plays a critical role in culture establishment because explant contamination during tissue culturing is extremely problematic. Explant contamination is most severe in woody species, which include all Amazonian fruit species. The potential for contamination is greatest when the explants are taken directly from the field. Although plants maintained under greenhouse conditions are easily controlled, explants derived from them also have a high potential for contamination. However, even plants maintained in nurseries or greenhouses and subjected to rigorous phytosanitary control can harbor microorganisms, which may limit *in vitro* culture procedures [20]. In [10] cautions that, even before harvesting material from the field, it is essential to take some measures for each step in the culture process, from the laboratory to acclimatization and plant development in greenhouses. The phytosanitary conditions of the mother plant help determine the ease of explant sterilization during isolation.

Several substances have been tested to minimize contamination and facilitate *in vitro* establishment, and chlorine and ethanol-based compounds are now standard disinfection tools (Silva et al., 2005). In some cases, antibiotics are added to the culture medium [29]. Research has demonstrated the success of these active ingredients in disinfecting explants of native fruit tree species when establishing plant tissues *in vitro* for use in studies on reproduction. The active ingredients in disinfectant solutions and their exposure durations vary greatly.

In studying decontamination of floral explants from cupuaçu, [29] recommended immersing the explants in a 0.25% sodium hypochlorite solution for 20 minutes. The authors also observed that adding cefatoxime antibiotic to the culture medium was vital in controlling contamination in these explants.

In studying decontamination of explants from bacuri trees, [29] and his team found that pre-treating the explants in an antifungal solution of carboxin (0.067% p/v), thiram (0.067% w/v), carbendazim (0.17% w/v), chlorothalonil (0.17% w/v), and thiophanate-methyl (0.067% w/v) [15], and immersing them in a 1.75% sodium hypochlorite solution for 30 minutes produced the best results.

In [31] found that the most efficient method for disinfecting the apices of peach palms was to immerse the peach palm explants in a 1.25% sodium hypochlorite solution for 20 minutes. The authors reported that 90% of these explants avoided contamination during *in vitro* culture. In [33] obtained the most promising results for the disinfection of explants from peach palm shoots; he first immersed the shoots in a 50% ethanol solution for 1 minute, and then in a 0.5% sodium hypochlorite solution for 5 minutes.

In Figure 1 you can see the success obtained in vitro establishment of camu-camu (*Myrciaria dubious*) conducted by staff of the Tissue Culture and Fruits Embrapa Roraima and UFRR.

In [34] and [35] note that explant oxidation during the *in vitro* establishment of native plants is common; in camu-camu, explant oxidation is a major obstacle to *in vitro* establishment. In [23] evaluated the efficiency of different concentrations of sodium hypochlorite solutions at different immersion durations in the disinfection of nodal segments of camu-camu. The authors observed that the lowest rate of contamination occurred when the explants were immersed in a 0.5% sodium hypochlorite solution for 10 minutes, but the rate of oxidation in the tissues was high for all treatments. Similarly, [21] found that immersing the explants in a 2% sodium hypochlorite solution for 10 to 20 minutes provided the best rate of disinfection in camu-camu seeds. However, these treatments killed the seeds.

Figure 1. *In vitro* establishment of camu-camu held in Biofactory (UFRR) by the team of Fruits and Tissue Culture of Embrapa Roraima and UFRR. Boa Vista, RR, 2010.

In evaluating the effect of a light regime on the mother plant and the branch type used for the *in vitro* establishment of araça (*Psidium* spp.), [13] observed lower rates of contamination when using nodal segment explants of herbaceous branches, and obtained the highest survival rates when the mother plants were kept in darkness. These results are likely due to the reduced exposure of younger branches to contamination in the environment from which the explants were harvested.

Another way to control explant oxidation during *in vitro* establishment is to reduce the concentrations of some components of the culture medium. In [33] found that reducing the concentrations of NH_4NO_3 and KNO_3 macronutrients, sucrose, and agar by 50% in MS culture medium [36] controlled phenolic oxidation and the proliferation of peach palm explant shoots.

During the reproductive stage of shoots *in vitro*, the culture medium's composition varies with the species and the type of tissue or organ used as an explant. MS is most commonly used in the micropropagation of plant species. However, the Woody Plant Medium (WPM) [37] has been used successfully in woody species, including native Amazonian fruit tree species.

When using growth regulators to induce new shoot growth, native plants generally respond better to a culture medium supplemented with low doses of auxins and/or cytokinins. In [38] obtained excellent results in nance trees using different benzyladenine (BAP) doses for micropropagation. However, high BAP concentrations (above 4.0 mg L^{-1}) were inefficient in inducing the growth of axillary shoots in nodal segments. In [30] found that BAP reduced the number of shoots and leaves in Inga tree explants.

4. Somatic embryogenesis

In somatic embryogenesis, somatic cells or tissues to develop the formation of a complete plant, through a series of stages similar to zygotic embryogenesis.

Most systems via somatic embryogenesis is indirect, in which somatic embryogenesis is induced and maintained through the multiplication. The advantage of this method is that large quantities of somatic embryos can be formed with minimal manipulation and laboratory space.

This process of plant regeneration was successfully achieved in cocoa [16, 17], cupuaçu [26,27] and murici [30].

In indirect somatic embryogenesis, the formation of pro-embryogenic callus is the first step in obtaining somatic embryos and is usually obtained with the culture of embryonic or juvenile explants in medium supplemented with auxin, especially 2,4-D and TDZ [39].

In [31] can be observed callus induction in shoot apices of peach palm. The highest percentage of induction was 60%, obtained by combining 10.0 mg L^{-1} 2,4-D and 3.0 mg L^{-1} BAP.

Some studies have shown the ability of different cupuaçu explants to form callus. In [27] studied the induction of somatic embryogenesis in explants of cupuaçu, concluded that the hypocotyl region proved to be the most responsive of the embryonic axis, forming callus-looking white and bright. The MS medium supplemented with 2,4-D promoted the formation of large callus. Similar results were found by [24], studying the effect of auxin concentration and liquid medium on the development of calluses cupuaçu. The authors observed that the combination of NAA and 2,4-D induced callus formation and root formation in hypocotyl segments, and coconut water in medium without growth regulators favored the rooting and callus formation.

In [40] also studied the induction of somatic embryogenesis in explants cupuaçu and observed that the callus appeared as a process influenced by genotype. The staminode is shown as a great source of explants for obtaining callus, and the PVP the best alternative to control the oxidation of explants cupuaçu. On the other hand, [41] did not succeed in callus formation using nodal segments cupuaçu in medium supplemented with different concentrations of 2,4 D. Similar results were obtained by [28]. The authors assessed the responses of different morphogenetic cupuaçu explants subjected to various culture conditions in vitro and argue that the absence of induction of somatic embryogenesis observed in culture may be related to several factors such as type and stage of development of the explants, using culture type and concentration of plant growth regulators.

In [25] undertook a study of the induction of callus formation of the hybrid *Theobroma grandiflorum* x *T.obovatum*, which has resistance to disease witches' broom (*Crinipellis perniciosa*), a disease that greatly affects native Amazonian species of the genus Theobroma. Among the explants, the cotyledons were produced more callus tissues in culture medium.

For açaizeiro, [12] reported the conversion can be in vitro isolated zygotic embryos from mature seeds and seedlings complete normal requiring the presence of NAA and BA in the culture medium. The concentration of 2.68 mM NAA combined with 1.11, 1.55 or 2.22 mM BAP promoted the best growth of the seedling shoot. The authors also found that required the presence of NAA and BA in the culture medium for the conversion of zygotic embryos and the early growth of seedlings grown in vitro. The same authors in studies on direct somatic embryogenesis using zygotic embryos of açai, satisfactory results, with a halving of the amount of nutrients in MS medium in the absence of fitoregulators.

5. *In vitro* zygotic embryo culture

In vitro plants require a source of exogenous energy for carrying out the photosynthesis. Sucrose has been used more carbon source being present in fruit culture media at concentrations ranging native 20-40 g L^{-1} [26]. In [9] found that immature embryos murmuru showed higher germination in medium supplemented with 30 g L^{-1} sucrose. For embryos obtained from ripe fruits, 15 g L^{-1} sucrose in the medium was sufficient for it to reach the best germination rates. These results confirm the hypothesis that depending on the species and stage of development of embryos, the presence of carbohydrate in the medium may be in minute concentrations, or even not necessary because of embryos of many species use the energy required for germination in vitro from their own reserves of the embryo [42].

Embryos younger typically require higher concentrations of carbohydrates in the culture medium to sustain germination [43, 44]. In [9] have obtained best results using 3% sucrose in the development of embryonic axes cupuaçu, in accordance with [16, 18, 19 and 45] in studies performed with *Theobroma cacao*.

6. *In vitro* germination of pollen grains

One of the techniques used to obtain new varieties is controlled hybridization in the field and subsequent evaluation of the progenies. Thus, to obtain success in breeding is important to catch up before going to the field the viability of pollen grains [46]. Thus, for the application of artificial pollination techniques, knowledge of the time of pollen viability and stigma receptivity is crucial for a successful fertilization of the flower [47]. However, the germination of the pollen grain depends on several factors, such as osmotic pressure, concentration and type of sugar, consistency, temperature, humidity, presence of enzymes and phytohormones in the middle [48, 46]. In native fruits, especially those species whose domestication process is more advanced, there are few studies in this line of research.

In [49] Oliveira et al. (2001), studying the viability of pollen *in vivo* and *in vitro* staining method in genotypes of açai, the success observed in *in vitro* pollen germination of different cultivars. The authors found that pollen grains in vivo assai, drawn from floral buds and newly opened, exhibit high viability, being higher in the second stage of evaluation. With respect to time of storage at -100ºC, in vitro pollen showed a reduction in viability with increasing storage time. We also found that the viability of pollen grains varied with genotype and stage evaluated. With this study, the authors still recommend that the pollen of genotypes can be used for controlled pollination without fertilization damages, including, may remain stored for up to one year of storage under the conditions tested in that work.

In [50] tested four concentrations of galactose, glucose, lactose and sucrose, with or without boric acid, for the germination of pollen from cubiuzeiro (*Solanum topiro* Humb. & Bonpl.) and cupuaçu (*Theobroma grandiflorum* Willd. Ex. Spreng. Schummann). The authors found that higher germination rates occurred in 10, 15 and 20% sucrose after 25 hours, no effect of boric acid for the two species.

In [48] in their studies to obtain high levels of pollen from *Theobroma grandiflorum* found that pollen from one flower is not uniform. Therefore, it is recommended to use all anthers of a flower to prepare a sample. It is not necessary to have more than 300 pollen grains per sample. The stage at which the button is to be collected for the purpose of artificial pollination or germination and when all are detached sepals, petals with the extended length, style, stigma and exposed about two hours after the onset of anthesis (Step E) The pollen collected should be used, preferably up to two hours after it is collected, where there was more viable. The author also recommends the optimal medium for pollen germination cupuaçu should consist of 5% lactose, 0.01 H_3BO_3 and 1% agar, pH 6.1. The pollen collected cupuaçuazeiro button at the stadium and can remain viable up to 72 hours after the plant collected and stored at ambient conditions on the button, but at the end of this period, viability is very low, about 5%.

7. Possible tissue culture techniques for native Amazonian fruit trees

There is a myriad of tissue culture techniques for native Amazonian fruit trees. However, this important technology has been seldom used.

Tissue culture techniques can be used for *in vitro* conservation, reproduction, exchange, and genetic resource conservation. These techniques are particularly useful in species that have recalcitrant seeds, low germination potential, and exclusively vegetative propagation, as well as in endangered species [51, 52].

Ovary cultures are important in selective plant breeding. Obtaining plants from haploid tissues and subsequently doubling the chromosome number through colchicine treatment produces homozygous strains quickly, eliminating the multiple generations of selfing required in conventional selective breeding.

Producing synthetic seeds from somatic embryos offers various advantages for propagating native fruit tree species, including year-round production, while avoiding the risk of losses from adverse weather, biomass degradation, pests, disease, and low production years. Moreover, synthetic seed technology enables more secure maintenance of the clonal identity of the material under laboratory conditions. Synthetic seeds can be sown directly into the field, thus eliminating the need for acclimatization structures, sowing, and nurseries.

Protoplast fusion in somatic plant cells has great potential for combining genomes from sexually incompatible species. Somatic hybrids (artificial polyploids) of different species or genera are obtained through protoplast fusion and the subsequent regeneration of plants. These polyploids can be used as rootstocks in fruit production.

8. Final thoughts

Although there are eminent importance of the native fruits of the Amazon, it appears that only a few studies with applications of tissue culture techniques in the domestication and breeding of these species. The works are practically concentrated in developing protocols for the establishment of *in vitro* plants, plus a few studies with somatic embryogenesis, immature embryo rescue protocols and studies of the germination of pollen grains to support breeding programs.

Despite the reality and effectiveness of the techniques of tissue culture, it is important that some basic difficulties must be overcome to implement satisfactory in woody fruit trees, as is the case of native Amazon. The phenolic oxidation disinfestation and represent the most serious problems during the establishment of in vitro culture of explants woody species.

Suitable types of explants removed at appropriate times can help by having a lower content of endogenous phenols in tissues and, thus, less oxidation. Likewise, minor physical and chemical damage at the time of excision and pest may help reduce the problem of oxidation. Furthermore, the addition of antioxidant compounds as cysteine, ascorbic acid and adsorbents such as activated carbon and PVP can be critical in preventing oxidation, which is more pronounced in the initial stages of cultivation [53]. Additonally, with cultivation at low light intensities, and frequent replacement of the culture medium, the chances of success in the establishment and cultivation of explants of woody species are quite high.

Another important aspect to be highlighted concerns the small number of fruit species native to the Amazon that has been the subject of study and application of techniques of tissue culture. These species can be summed up pretty much to those whose breeding programs are in advanced stage. Thus, it appears that there is still vast field of study to be explored and a significant deficiency of techniques and information that will enable progress in the domestication and breeding of these species.

9. Conclusion

Although native Amazonian fruit tree species are important, few studies on the use of tissue culture techniques for domesticating and improving these species exist. Research has instead focused on developing protocols for the *in vitro* establishment of plants, and some studies have examined somatic embryogenesis, immature embryo rescue, and germination protocols for pollen grains as a basis for selective breeding programs.

Author details

Moacir Pasqual, Joyce Dória Rodrigues Soares and Filipe Almendagna Rodrigues
Federal University of Lavras (UFLA), Department of Agriculture, Lavras/MG, Brasil

Edvan Alves Chagas
Brazilian Agricultural Research Corporation (EMBRAPA), Distrito industrial, Boa Vista-RR, Brasil

10. References

[1] Fernandes ARF, Carvalho J G, Melo PC. Efeito do fósforo e do zinco sobre o crescimento de mudas do cupuaçuzeiro (*Theobroma Grandiflorum Schum.*) Cerne 2003; 9(2):221-230.

[2] Franzon RC. Caracterização de mirtáceas nativas do sul do Brasil. Dissertation. Universidade Federal de Pelotas; 2004.

[3] Auler AS, Wang X, Edwards RL, Cheng H, Cristalli OS, Smart PL, Richards DA. Palaeoenvironments in semi-arid northeastern Brazil inferred from high precision mass spectrometric speleothem and travertine ages and the dynamics of South American rainforests. Speleogenesis and Evolution of Karst Aquifers 2004; 2:1-4.

[4] Rebouças FS, Carmo DO, Almeirda NS, Dantas ACVL, Costa MAPC, Almeirda WAB. Germinação de cupuaçuzeiro *in vivo* e *in vitro*. In: XX Congresso Brasileiro de Fruticultura 54th Annual Meeting of the Interamerican Society for Tropical Horticulture, 12-17 October 2008, Vitória; 2008.

[5] Mitra SK. Important *Myrtaceae* Fruit Crops. Acta Horticulturae 2010; 849:33-38.

[6] Anuário brasileiro de fruticultura. Panorama. Editora Gazeta; 2008.

[7] Luis ZG. Propagação in vitro e caracterização anatômica de gemas adventíceas e embriões somáticos de murici (Byrsonima basiloba Juss., Malpighiaceae). Dissertation. Universidade de Brasília; 2008.

[8] Sudam DRN, Genamaz CT. Workshop biodiversidade de fruteiras com potencial sócio-econômico na Amazônia: 24-26 August 1999; relatório final. Belém; 2000.

[9] Pereira JES, Maciel TMS, Costa FHS, Pereira MAA. Germinação in vitro de embriões zigóticos de murmuru (Astrocaryum ulei). Ciência Agrotécnica 2006; 30(2): 251-256.

[10] Pasqual, M. Introdução à cultura de tecidos: fundamentos básicos. Lavras: UFLA/FAEPE; 2001.

[11] Borém, A. Biotecnologia Aplicada ao Melhoramento de Espécies Silvestres. In: BORÉM, A.; LOPES, M. T. G.; CLEMENT, C. R. Domesticação e Melhoramento de Espécies Amazônicas. Ed. UFV. Viçosa: UFV; 2009, p117-163.

[12] Ledo AS, Lameira AO, Benbadis AK, Menezes IC, Ledo CAS, Oliveira MSP. Cultura in vitro de embriões zigóticos de açaizeiro. Revista Brasileira de Fruticultura 2001; 23(3):468-472.

[13] Souza JÁ. Efeito do tipo de ramo e do regime de luz fornecido à planta matriz no estabelecimento in vitro de araçazeiro cv. "Irapuã". Ciência Rural 2006; 36(6):1920-1922.

[14] Rodrigues EF. Desenvolvimento do eixo embrionário in vitro e calogênese de cupuaçu (Theobroma grandiflorum (Willd. ex Spreng.) Schum.) e estabelecimento do ápice caulinar de bacuri (Platonia insignis Martius). Thesis. Universidade Estadual Paulista; 2000.

[15] Ferreira MGR, Santos MRA, Santos ER, Rocha JF, Correia AO. Desinfestação de explantes radiculares de bacurizeiro (Platonia insignis Mart.). Saber Científico 2009b; 2(2):56-62.

[16] Duhem K, Le Mercier N. Données nouvelles surl' induction et le développement d' embryons somatiques chez Theobroma cacao L. Café Cacao Thé 1989; 33(1):9-14.

[17] Söndahl MR, Liu S, Bellato C, Bragin A. Cacao somatic embryogenesis. Acta Horticulturae 1993; 336:245-248.

[18] Janick J, Whipkey A. Axillary proliferation of shoots from cotyledonary nodal tissue of cacao. Revista Theobroma 1985; 15:125-131.

[19] Kononowicz AK, Janick J. In vitro development of zygotic embryos of Theobroma cacao. Journal of the American Society of Horticultural Science 1984; 109(2):266-269.

[20] Medeiros CPC. Indução in vitro de respostas morfogenéticas em explantes nodais de cajazeira (Spondias mombin L.). Dissertation, Universidade Federal do Ceará; 1999.

[21] Kikuchi TYP, Nunes HCB, Mota MGC, Vieira IMS, Ribeiro SI, Corrêa MLP. Assepsia para sementes de camu-camu (Myrciaria dubia (H.B.K) Mc vaugh) cultivadas in vitro. In: Anais do IV Encuentro Latinoamericano de Biotecnologia Vegetal; 2001.

[22] Gutiérrez-Rosati A, Rojas EI, Micky M, Rodriguez M. Avances em la introducción de genotipos de camu camu (*Myrciaria dubia* H.B.K. Mc Vaugh). Folia Amazônica 2006; p.57-60,

[23] Araújo MCR, Chagas EA, Couceiro MA, Donini LP, Pio R, Schwengber JAM, Castro AM, Araújo WF. Desinfestação *in vitro* de segmentos nodais de camu-camu. in: XXI Congresso Brasileiro de Fruticultura, Natal; 2010.

[24] Ferreira MGR, Cárdenas FEN, Carvalho CHS, Carneiro AA, Damião Filho CF. Desenvolvimento de calos em explantes de cupuaçuzeiro (*Theobroma grandiflorum* Schum) em função da concentração de auxinas e do meio líquido. Revista Brasileira de Fruticultura 2001; 23(1):473-476.

[25] Venturieri GA, Venturieri GC. Calogênese do híbrido *Theobroma grandiflorum* x *T.obovatum* (Sterculiaceae). Acta Amazônica 2004; 34(4):507–511.

[26] Ferreira MGR, Cárdenas FHN, Carvalho CHSC, Carneiro AA, Dantas Filho CF. Resposta de eixos embrionários de cupuaçu (*Theobroma grandiflorum* Schum.) à concentração de sais, doses de sacarose e renovação do meio de cultivo. Revista Brasileira de Fruticultura 2002; 24(1):246-248.

[27] Ferreira MGR, Cárdenas FEN, Carvalho CHS, Carneiro AA, Filho CFD. Indução de calos embriogênicos em explantes de cupuaçuzeiro. Revista Brasileira de Fruticultura 2006; 26(2):372-374.

[28] Ledo AS, Lameira AO, Benbadis AK. Explantes de cupuaçuzeiro submetidos a diferentes condições de cultura *in vitro*. Revista Brasileira de Fruticultura 2002. 24(3):604-607.

[29] Ferreira MGR, Santos MRA, Bragado ACR. Propagação *in vitro* de cupuaçuzeiro: desinfestação de explantes florais. Saber Científico 2009a; 2(2):37-44.

[30] Stein VC. Organogênese direta em explantes de ingazeiro (*Inga vera* Willd. subsp. *affinis* (DC.) T.D. Penn.). Revista Brasileira de Biociências 2007; 5(2):723-725.

[31] Santos MRA, Ferreira MGR, Correira AOC, Rocha JF. *In vitro* establishment and callogenesis in shoot tips of peach palm. Revista Caatinga 2010; 23(1):40-44.

[32] Bastos LP. Cultivo *in vitro* de mangabeira (*Hancornia speciosa*). Revista Brasileira de Biociências 2007; 5(2):1122-1124.

[33] Teixeira JB. Limitações ao processo de cultivo *in vitro* de espécies lenhosas. Brasília: Embrapa-Recursos Genéticos e Biotecnologia; 2001.

[34] Sato AY, Dias HCT, Andrade LA, Souza VC. Micropropagação de Celtis sp: controle da cantaminação e oxidação. Cerne 2001; 7(2):117–123.

[35] Erig CA, Schuch MW. Estabelecimento *in vitro* de plantas de marmeleiro (*Cydonia oblonga* Mill.) cultivares MC, Adans e Portugal. Ciência Rural 2003; 8(2):107-115.

[36] Murashige T, Skoog FA. revised medium for rapid growth and bioassays with tabacco tissue culture. Physiologia Plantarum 1962; 15:473-497.

[37] Lloyd G, Mccown B. Commercially feasible micropropagation of montain laurel, *Kalmia latifolia*, by use of shoot tip culture. Com. Proc. Int. Plant Prop. Soc. 1981; 30:421-327.

[38] Nogueira RC. Indução de calos em explantes foliares de murici-pequeno. Ciência e Agrotecnologia 2007; 31(2):366-370.

[39] Akram M, Aftab F. High frequency multiple shoot formation from nodal explants of teak (*Tectona grandis* L.) induced by thidiazuron. Propagation of Ornamental Plants 2008; 8(2):72-75.

[40] Almeida CF, Rodrigues SM, Lemos OF. INDUÇÃO DE CALOS EMBRIOGÊNICOS EM EXPLANTES DE CUPUAÇU. In: XIV Seminário de Iniciação Científica da EMBRAPA. Embrapa Amazônia Oriental, Belém; 2010.

[41] Roberts SA, Cameron RE. The effects of concentration and sodium hydroxide on the rheological properties of potato starch gelatinization. Carbohydrate Polymers 2002; 50: 2446-2455.

[42] García JL, Troncoso J, Sarmiento R, Troncoso A. Influence of carbon source and concentration on the in vitro development of olive zygotic embryos and explants raised from them. Plant Cell, Tissue and Organ Culture, Dordrecht, 69:95-100, 2002.

[43] Chagas EA, Pasqual M, Dutra LF, Silva AB, Cazetta JO, Santos FC, Cardoso P. Desempenho de diferentes estádios embrionários no cultivo in vitro de embriões de 'Pêra Rio' x 'Poncã'. Revista Brasileira de Fruticultura 2003a; 25(3): 523-525.

[44] Chagas EA, Pasqual M, Ramos JD, Cardoso P, Cazetta JO, Figueiredo MA. Development of globular embryos from the hybridization between 'Pêra Rio' sweet orange and 'Poncã' mandarin. Revista Brasileira de Fruticultura 2003b; 25(3):483-488.

[45] Söndahl MR, Liu S, Bellato C, Bragin A Cacao somatic embryogenesis. Acta Horticulturae, Wageningen, 336:245-248, 1993.

[46] Chagas EA, Pio R, Chagas PC, Pasqual M, Neto JEB. Composição do meio de cultura e condições ambientais para a germinação de porta-enxertos de pereira. Ciência Rural 2010; 40: 261-266.

[47] Ramos JD, Pasqual M, Pio LAS, Chagas EA, Pio R. Stigma receptivity and in vitro citrus pollen grains germination protocol and adjustment. Interciência 2008; 33:51-55.

[48] Antonio IC. *In vitro* germination of cupuassu [Theobroma grandiflorum (Willdenow ex Sprengel) Schumann] pollen grain. Científica 2004; 32(2):101-106.

[49] Oliveira MSP, Maués MM, Kalume MAA. Viabilidade de pólen in vivo e in vitro em genótipos de açaizeiro. Acta Botânica Brasileira 2001; 15(1):27-33.

[50] Neves TS, Machado GME, Oliveira RP. Efeito do tipo e concentração de carboidratos e ácido bórico na germinação de grãos de pólen de cubiuzeiro e cupuaçuzeiro. In: Congresso brasileiro de fruticultura, 14.; reunião interamericana de horticultura tropical, 42.; Simpósio internacional de mirtáceas, 1., 1996, Curitiba. Resumos... Curitiba: Sociedade Brasileira de Fruticultura, 1996. p.213.

[51] Andrade SEM. Princípios da cultura de tecidos vegetais. Planaltina: Embrapa Cerrados; 2002.

[52] Engelmann F, Engels JMM. Technologies and Strategies for *ex situ* conservation. In: ENGELS, J. M. M. et al. (Eds.). Managing Plant Genetic Diversity. Oxford: IPGRI; 2002, p89-103.

[53] Santos RB, Paiva R, Paiva PDO, Santana JRF Problemas no cultivo *in vitro*: cultura de tecidos. Paiva e Paiva, UFLA, Lavras, MG. 9:73-79, 2001.

Chapter 10

Effect of Additives on Micropropagation of an Endangered Medicinal Tree *Oroxylum indicum* L. Vent

Y.K. Bansal and Mamta Gokhale

Additional information is available at the end of the chapter

1. Introduction

Sonpatha (*Oroxylum indicum* (L.) Vent.) is a threatened medicinal tree species [1,2] belonging to family Bignoniaceae. It is valued for its antimicrobial, antiarthritic, anticancerous and antihepatitic properties possessed by its various parts. Root extract of this tree has been used for long in ayurvedic preparations like Dashmularisht and Chyawanprash [3,4].This tree possesses a flavonoid viz. Baicalein used to check proliferation of human breast cancer cell line MDA - MB - 435 [5]. Sonpatha grows in India, Sri Lanka, South China, Celebes, Philippines and Malaysia[6,7]. In India, it is distributed throughout the country up to an altitude of 1200 m and found mainly in ravine and moist places in the forests [8].Owing to indiscriminate collection, over exploitation and uprooting of whole plants with roots, this valuable tree has become vulnerable in different states of India like Karnataka, Andhra Pradesh, Kerala, Maharastra, M.P. and Chhatisgarh [9,10]. Hence research towards mass multiplication, conservation and higher production of the active compound under *in vitro* culture conditions is essential [11]. Few reports are available on the *in vitro* regeneration of the species [12,13]. Optimum factors influencing growth and morphogenesis vary with the genotype and types of explants used for micropropagation. [14]. Murashige and Skoog (MS) medium with a high content of nitrate, ammonium and potassium is of widespread use in the successful culture of a wide variety of plants. Sometimes it requires supplementation of additional substances in the medium.

Application of additives is adapted to the cultural needs[15] i.e. objectives of the experimental studies like micropropagation, regeneration, cytodifferentiation, androgenesis, biosynthesis of secondary metabolites and biotransformation of cells as well as the particular plant species taken. In this chapter the importance of some additives like activated charcoal (AC), casein hydrolysate (CH), coconut milk (CM) & silver nitrate (AgNO₃) & their impact on the direct & indirect *in vitro* multiplication of a threatened medicinal tree Sonpatha (*Oroxylum indicum*) is emphasized.

1.1. Activated charcoal (AC)

Activated charcoal (Carbonized wood) is a fine powdered wood charcoal added to tissue culture media, to bring about changes in the composition of the medium[16]. Being porous, it serves to adsorb toxic & phenolic tissue exudates in culture, which prevents inhibition of growth, promotes embryogenesis/ organogenesis Beneficial effects of addition of activated charcoal to media are highlighted by various researchers [17,18].

Activated charcoal (AC) has a very fine network of pores with large inner surface area on which many substances can be adsorbed & it is often used in tissue culture to improve cell growth and development. It plays critical roles in micropropagation, seed germination, somatic embryogenesis, anther culture, synthetic seed production, protoplast culture, rooting, stem elongation and bulb formation in different plants. The beneficial effects of AC on morphogenesis may be mainly due to its irreversible adsorption of inhibitory compounds in the culture medium and substantially decreasing the toxic metabolites, phenolic exudation and brown exudate accumulation [19]. In addition to this activated charcoal is involved in a number of stimulatory and inhibitory activities including the release of substances naturally present in AC which promote growth and darkening of culture media, adsorption of vitamins, metal ions and plant growth regulators, including abscisic acid and gaseous ethylene [20]. The effect of AC on growth regulator uptake is still unclear but some workers believe that AC may gradually release certain adsorbed products, such as nutrients and growth regulators which become available to plants. This review focuses on various roles of activated charcoal in plant tissue culture and the recent developments in this area.

1.2. Coconut milk (CM)

A natural complex may be used when a defined medium fails to support a particular growth response. Its addition makes a defined medium undefined since variations are to be expected in growth promoting or inhibitory compounds in these complexes [21]. A liquid endosperm such as coconut milk would be a good medium for embryo culture. It was first used successfully for culture of very young embryos of *Datura*[22]. Explants proliferate more readily on CM containing media than that observed with auxin. It enhances the proliferation of tumoral tissues indicating that it contains a stimulating substance different from an auxin. Addition of coconut milk serves to rejuvenate mature and permanent cells into actively dividing cells, promoting cell division & callus formation [23]. The composition of CM has been investigated extensively [24] but the analysis has been complicated by the variability in age of coconuts from which the liquid endosperm was obtained.

1.3. Casein hydrolysate (CH)

Casein hydrolysate (CH) is an organic nitrogen supplement containing a mixture of amino acids. Being a good source of reduced nitrogen it has been widely used as an additive to embryo culture media [25,26]. It has proved superior to the combined effect of the amino acid mixture. It has been thought that nitrogen deficiency can cheaply be fulfilled by its addition [27,28] presumably it contains some stimulatory factors yet unidentified.

1.4. Silver nitrate (AgNO₃)

Effect of nitrate supplementation in media has been well established in tissue culture [29,30] so as to enhance shoot multiplication and somatic embryogenesis. Several researchers assume that NO_3 /NH_4 $^+$ ratio acts as buffering stabilization of medium pH resulting *in vitro* organization by adventitious shoot (apical meristem) and NO_3 subsequently promotes extension growth by these meristems into full fledged shoots.

Silver nitrate works as an inhibitor of ethylene activity [31] through the Ag^{2+} ions by reducing the receptor capacity to bind ethylene [32-35].With these observations, water solubility and lack of phytotoxicity at effective concentration led to its application in tissue culture [36].

2. Materials and methods

2.1. Plant materials and culture conditions

Seeds of *O. indicum* were collected from forest areas in and around Jabalpur (M.P.), India. Seeds were germinated on moist sterilized filter paper under *in vitro* conditions. Fifteen to twenty days old seedlings were given a treatment of 1 minute each of 70% ethyl alcohol and 0.1% mercuric chloride followed by sterilized water washing (3-4 times) and excess water was blotted on sterile filter paper. The explants viz. apical buds (ApB) (0.5-1cm) , axillary buds (AxB) (0.7-1cm) and embryonic axis (Ea) (0.4-0.6 cm) were excised and explants were inoculated under aseptic conditions in test tubes on Murashige and Skoog's (MS) medium [37] supplemented with 3% sucrose, 0.8 % agar and different concentrations (0.1-10 mgL⁻¹) of plant growth regulators viz. auxins (2,4-Dichloro phenoxy acetic acid (2,4-D), Indole butyric acid (IBA), Naphthalene acetic acid (NAA) and Indole acetic acid (IAA)) individually. The pH of the media was adjusted to 5.7 before adding agar. Medium (8-10 ml) was dispensed in glass test tubes (15x125 mm) and autoclaved at a pressure of 15 psi and a temperature of 121ºC for 15 minutes. Before inoculation autoclaved medium was left at 25ºC for 24 hrs to check that there was no visible microbial contamination. A piece of callus (2-3mm x 2-3 mm) raised on auxin was subsequently used for indirect organogenesis. The cultures were maintained in culture room at a temperature of 25±2ºC, relative humidity (RH) of 60-70% and a light intensity of approx. 1500 lux provided by cool, white, fluorescent tubes under a photoperiod of 16/8 hr (light/dark).

2.2. Callus induction, plant regeneration and rooting

Apical bud, axillary bud and embryonic axis explants were inoculated on MS medium supplemented with different concentrations of plant growth regulators to induce multiple shoots & callus. Calli were subcultured onto fresh medium every 20-22 days for further proliferation on suitable medium. Regenerated shoots were elongated up to approx 2 cm, excised and transferred to MS medium fortified with different (0.1, 0.5, 1.0, 5.0 mgL⁻¹) concentrations of NAA, IBA and IAA for root induction.

2.3. Fortification of additives into the media

Out of different plant growth regulators (auxins and cytokinins) used , the frequency of shoot initiation, rate of multiplication and shoot length was significantly high on BAP (1mgL-1) supplemented media (Selected medium SM) in both *in vitro* regeneration routes (both directly i.e. from explants ApB, AxB & Ea as well as indirectly from callus. Addition of different concentrations of additives (CH, CM, AC & AgNO₃) was studied on different explants & calli in SM. Most of the cultures have been established to study their morphogenic potential up to three subcultures. With this view the shoot buds regenerated indirectly were multiplied on the same concentration of PGR on which shoots got initiated up to 3 subcultures (one sub culture Passage of 20-22 days). Although the three auxins (IBA, NAA, IAA) induced roots in *in vitro* raised shoots of *Oroxylum indicum* IBA turned out to be the best for all parameters of rooting. Shoots with the highest frequency of root induction and maximum number of elongated roots were developed on MS medium containing IBA (1mg L-1). MS medium with IBA (1mg L-1) was selected to apply additives for further rooting experiments (data are not shown for effect of CM & CH on rooting).

2.4. Hardening and acclimatization

Approximately four-month-old plants bearing a well-developed root system were washed carefully to remove traces of agar. The plantlets (5 month old) remained fresh when transferred to conical flask with root system immersed in distilled water (4 days) followed by ordinary water (4 days). Such vigorously growing regenerated plantlets were then transferred to pots containing soil: sand (1:1) mixture for 15 days. Approx 82 % of the hardened plants survived in the pot.

2.5. Data recording:

To test the efficiency of direct shoot regeneration, frequency of shoot induction (FSI) directly from different explants was noted. While for indirect shoot regeneration, frequency of shoot regenerated from callus was calculated. Same parameters were calculated for rooting. The effect of continuous supplementation of plant growth regulators on indirect shoot regeneration was observed up to three subculture passages each of 20-22 days. Shoot buds obtained from I subculture passage were subsequently used as explants for II and III subculture passages. All experiments were completely randomized and repeated twice. Each treatment consisted of 25 replicates.

3. Results and discussion

3.1. Activated charcoal

Activated charcoal has been reported to inhibit heavy leaching of phenolics [38]. In the present work AC successfully overcame this problem during regeneration resulting in shoots with good shoot length. However, the shoots formed in the presence of the AC were rather weak with small leaves (Plate 1, Figs. 1a-1c). Even at low concentration it (activated charcoal)

Plate 1. Effect of additives on multiple shoot regeneration from ApB and AxB explants of *O. indicum* (L.) Vent. (1a-1c: Effect of AC on multiple shoot regeneration from AxB explants, 1d-1e: Effect of AC on multiple shoot regeneration from ApB explants, 2a-2d: Effect of $AgNO_3$ on multiple shoot regeneration from AxB explants in I subculture passage 2e-2f II subculture passage, 2g-2h: III subculture passage, 3a-3b: Effect of $AgNO_3$ on multiple shoot regeneration from ApB explants, 4: Elongated shoot, 5: Regeneration from embryonic axis explant

inhibited multiple shoot formation (direct) (Graphs 1, 2) as well as callusing at the base of regenerated shoots. Inhibition of multiple shoot formation after AC supplementation was also observed in *Ficus carica* [39]. Rooting of *in vitro* regenerated shoots initiated after 20-25 days of inoculation on IBA (1mgL^{-1}) supplemented medium containing AC. Only low and moderate (2, 4 %) concs. of AC were effective, while high conc. (6%) failed to induce rooting. Roots developed in *in vitro* regenerated shoots on this rooting media were long, thin, unbranched and too weak to be transferred for hardening process.

3.2. Coconut milk

The explants remained totally unresponsive when cultured on MS medium fortified with CM. Significant reduction in frequency of shoot initiation, shoot number (SN) and shoot length (SL) was found. Mostly shoots developed with only one leaf at a node. Leaves were small in size and light green in colour (Plate 2, Figs. 6a-7b). Increment in basal callusing was observed after 6-8 days of inoculation. Callus was fresh, light brown in colour and non-regenerative in nature. Coconut milk supplementation as an additive in combination with BAP did not support multiplication of shoot either directly or indirectly in *O. indicum*. This observation supports the past reports [40, 41].

3.3. Casein hydrolysate (CH)

Casein hydrolysate (CH) supplementation to culture medium successfully overcame inhibition of regeneration from explants directly as well as indirectly. The number of shoots was found to be enhanced on CH (20 mgL^{-1}) supplemented medium resulting in 9.34 fold increase over control. Higher concs. of CH (30 and 40 mgL^{-1}) resulted in no further increase in the number of shoots (Graphs 1-3). CH has also been found useful in *Anogeissus pendula* & *A. latifolia* [42]. Induction of healthy shoot formation has been reported in *Crataeva nurvala* [43] using CH.

Compared to shoot number (SN) shoot length (SL) was adversely affected at all concs. of CH enhancing only 0.5 to 2 cm shoot length in apical meristem derived shoots and 0.7 –1.0 cm in axillary bud derived shoots respectively (Plate 2, Figs. 1a-5). CH (500 mg L^{-1}) supplemented medium did not support shoot growth as a consequence of which shoots remained compact and stunted. In some cases reduced concentrations of CH induced elongation of shoots [44]. CH was unable to induce indirect multiple shoot formation from calli as efficiently as directly from the explants (Graph 4).

3.4. Silver nitrate (AgNO₃)

Maximum frequency of shoot intiation has been observed on AgNO₃ supplemented medium in both types of shoot regeneration systems (direct and indirect) among all the additives attempted (Graph1). Explants when treated with different concs. of AgNO₃ (0.1, 1, 2, 4 mgL^{-1}) with BAP (1mgL^{-1}) resulted in the formation of healthy shoots bearing large dark green leaves (Plate 1, Figs. 2a-5). Silver nitrate has produced positive effect on all the shoot

Plate 2.   Effect of additives on multiple shoot regeneration from ApB and AxB explants of *O. indicum* (L.) Vent. 1a-2a: Effect of CH on multiple shoot regeneration from ApB explants, 2b-5: Effect of CH on multiple shoot regeneration from AxB explants , 6a-6b: Effect of CM on multiple shoot regeneration from AxB explants, 7a-7b: Effect of CM on multiple shoot regeneration from ApB explants.

Plate 3.   Effect of additives on multiple shoot regeneration (indirect) and rooting from in *O. indicum* (L.) Vent. 1a-2b: Effect of silver nitrate AgNO3 (2mgL⁻¹) on indirect multiple shoot formation (1a-b:I subculture passage, 1c-d: II subculture passage, 2a-2b: III subculture passage), 1-8: Rooting in *in vitro* regenerated shoots on IBA (1mgL⁻¹)+ AgNO3 (2mgL⁻¹) +MS medium

regeneration parameters FSI (Frequency of Shoot Initiation), MNS (Mean Number of Shoots), MSL (Mean Shoot Length) for direct and indirect regeneration, Graph 1-4) and rooting parameters FR (Frequency of Rooting), MNR (Mean Number of Roots), MRL (Mean Root Length) (Graph 5) tested for regeneration of *O. indicum* by developing healthy plantlets (Photoplate1:2a-5; Photoplate 3). Silver nitrate has been used previously to prevent callus formation in tree and woody species viz. *Garcinia mangostana* [45], *Albizia procera* [46] and *Manihot esculanta* [47] as well as in other plants viz. *Vanilla planifolia* [48]. In the present study multiple shoot proliferation & elongation of shoots (Table-1) were enhanced efficiently on selected medium (SM) by $AgNO_3$ (2 mgL^{-1}). Such silver nitrate supported multiple shoot formation has been reported in different plants viz. *Coffea* sp. [49, 50] and *Brassica* sp. [51, 52].Supplementation of $AgNO_3$ in culture media caused significant positive effect on shoot number with its best response being observed at 2 mgL^{-1} in subculture passage-III. In some plants, the regeneration potential of cultured cells and tissues has been reported to decrease with increasing cycle of subcultures [53]. Incorporation of $AgNO_3$ in the culture media in the present species enhances shoot multiplication up to sub culture passage III (Graph 4) by retaining the regeneration potential as reported in *Albizzia julibrissin* & *Nicotiana plumbaginifolia* [15].

Addition of $AgNO_3$ in culture media resulted in maximum rooting frequency and root length (Graph 5) of *in vitro* regenerated shoots in the present study. Silver nitrate induced rooting has been reported in *Vanilla* [48], *Decalepis hamiltonii* [54, 55] and *Rotula aquatica* Lour. [56].

4. Conclusion

The *in vitro* regeneration of some plants remains difficult due to high degree of callusing, high phenolic excretion into the medium and consequent blackening of explants. Fortification of culture media with different plant growth regulators i.e. auxins and cytokinins is not enough to regenerate the plant with high efficiency. This type of cultures in some cases may be improved by incorporation of additives in the media due to their growth and development promoting activities. In the present work additives used were Casein hydrolysate (CH), Activated Charcoal (AC), Coconut milk (CM) and Silver nitrate ($AgNO_3$) to induce *in vitro* regeneration of an important endangered medicinal tree species *Oroxylum indicum* (L.) Vent. Among all the additives used CH and $AgNO_3$ acted positively for multiple shoot regeneration from different explants (ApB, AxB and Ea) directly as well as indirectly by overcoming inhibition during regeneration. Whereas more conspicuous role of Casein hydrolysate (CH.) in *O. indicum* seems to be on the number of shoots induced while that of $AgNO_3$ was mainly on shoot lengths besides the number of shoots produced. Also $AgNO_3$ favors efficient rooting from *in vitro* regenerated shoots when supplemented in combination with auxin IBA. Overall Silver nitrate has turned out to be the best additive for regeneration of *O. indicum*. The production of secondary metabolites from *in vitro* regenerated vis-a-vis nature grown tissues is expected to provide useful information in future.

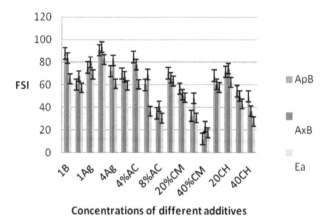

Figure 1. Effect of BAP +Additives on FSI from different explants of O. indicum (L.) Vent.

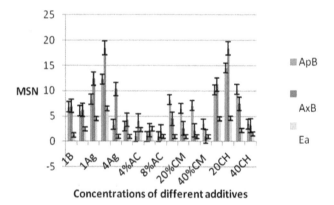

Figure 2. Effects of BAP +Additives on MSN from different explants of O.indicum Vent.

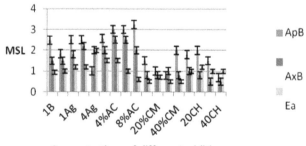

Figure 3. Effect of BAP +Additives on MSL from different explants of O. indicum (L.) Vent.

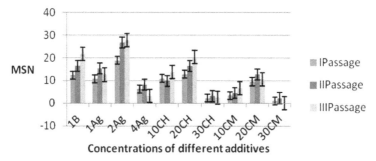

Figure 4. Effect of BAP +Additives on MSN (indirect) in O.indicum (L.) Vent

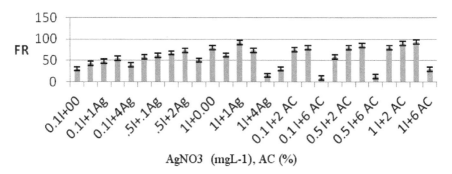

Figure 5. Effect of AC and AgNO₃ with IBA on rooting from *in vitro* regenerated shoots of *O. indicum*

Author details

Y.K. Bansal and Mamta Gokhale
Plant Tissue Culture Laboratory,
Department of Biological Science, R.D. University, Jabalpur-M.P. India

5. References

[1] Darshan S, Ved D K (2003) A Balanced Perspective for Management of Indian Medicinal Plants. Ind. forest. 275-288.

[2] Ved D K, Kinhal G A, Ravikumar K, Mohan K, Vijayshankar R, Indresha J H (2003) Threat Assessment and Management Prioritization FRLHT Bangalore 87-88.

[3] Chen L, Games D E, Jones J (2003) Isolation and Identification of Four Flavonoid Constituents from the Seeds of *Oroxylum indicum* by High Speed Counter Current Chromatography. J. chrom. 988(1): 95-105.

[4] Yasodha R, Ghosh M, Santan B, Gurumurthi K (2004) Importance of Biotechnological Research in Tree Species of Dashmula. Ind. forest. 130:79-88.

[5] Lambertini E., Piva R, Khan M T H, Bianchi N, Borgatti M, Gambari R (2004) Effects of Extracts from Bangladeshi Medicinal Plant *In Vitro* Proliferation of Human Breast Cancer Cell and Expression of Estrogen Receptor Alpha Gene. Int. j. oncology 24: 419-423.

[6] Anonymous (1966) Wealth of India. Vol. II, Publication and Information Directorate, CSIR, New Delhi, India 107-108,

[7] Dey A C (1980) Indian Medicinal Plants Used in Ayurvedic Preparations. Bishen Singh, Mahendra Pal Singh, Dehradun pp 202.

[8] Bennet S S R, Gupta P C, Rao R V (1992) Venerated Plants. ICFRE, Dehradun pp. 147-149.

[9] Jain S P, Singh J, Singh S C (2003) Rare and Endangered Medicinal and Aromatic Plants of Madhya Pradesh. J. econ. taxon. bot. 27:4:925-932.

[10] Jayaram K, Prasad M N V (2008) Genetic Diversity in *Oroxylum indicum* (L) Vent. (Bignoniaceae) A Vulnerable Medicinal Plant by Random Amplified Polymorphic DNA Marker Afr. j. biotechnol. 7(3): 254-262.

[11] Gokhale M, Bansal Y K (2006) An avowel of Importance of an Endangered Tree Shivnak (*Oroxylum indicum* (L) Vent). Nat. prod. rad. 5(2): 112-114.

[12] Gokhale M and Bansal Y K (2009) Direct *In Vitro* Regeneration of a Medicinal Tree *Oroxylum indicum* (L.) Vent. Through Tissue Culture. Afr. j.biotechnol. 8 (16): 3777-3781.

[13] Gokhale M and Bansal Y K (2010) Indirect organogenesis in *Oroxylum indicum* (L.) vent. Ind. Forest. 136(6): 804-811.

[14] Fossard R A, de Giladi I, Altman A, Goren R *(1977)* Tissue Culture in Horticulture- A Perspective. Acta hortic. 78: 455-459.

[15] Vinod K, Parvatam G, Gokare A R (2009) AgNO3 – A Potential Regulator of Ethylene Activity and Plant Growth Modulator. Electronic j. biotech.12 (2): 1-15.

[16] Van Winkle S C, Pullman G S (1995) The Role of Activated Carbon in Tissue Culture Medium Energeia CAER University of Kentucky, Center for Applied Energy Research.6(6): 2-4 .

[17] Groll J, Gray V M , Mycock D J (2002) Development of Cassava (*Manihot esculanta* Crantz.) Somatic Embryos During Culture with Abscisic Acid and Activated Charcoal J. plant physiol. 159 (4):437–443

[18] Thomas D (2008) The Role of Activated Charcoal In Plant Tissue Culture. Biotechnol. advan. 26(6): 618–631.

[19] Pan M J, van Stedan J (1998) The Use of Charcoal In *In Vitro* Culture- A Review. Plant growth regu. 26(3):155-163

[20] Peck D E, Cummings B G (1986) Beneficial Effects of Activated Charcoal on Bulblet Production In Tissue Cultures of *Muscari armeniacum*. Plant cell tiss. org. cult. 6:9-14

[21] Mishra S.P. (2009) (editor) Culture Media, Growth Regulators and Nutrient Supplements In: Plant Tissue Culture. Ane Books Pvt. Ltd pp 41-50

[22] van Overbeek J, Conklin M E, Blakeslee A F (1941) Factors in Coconut Milk Essential For Growth and Development of Very Young *Datura* Embryos. Science 94:350-351.

[23] Gautheret R J (1985) History Of Plant Tissue & Cell Culture: A Personal Account. In: Vasil I.K. editor, Cell Culture & Somatic Cell Genetics of Plants Vol. II Cell Growth, Nutrition, Cytodifferentiation & Cryopreservation. Academic Press, Inc. pp2-50.

[24] Yong J W H, Ge L, Ng Y F, Tan S N (2009) Chemical Composition and Biological Properties of Coconut (*Cocos nucifera* L.) Water. Molecules 14:5144-5164

[25] Cameron-Mills V, Duffus C M (1977) The *In Vitro* Culture of Immature Barley Embryos on Different Culture Media. Ann. bot. 41: 1117 - 1127.

[26] Narayanswamy S (2007) Tissue Nutrition –Growth Hormones. Plant Cell and Tissue Culture. Tata McGraw-Hill Publishing Company Ltd, New Delhi pp 22-50.

[27] George E F (1993) Plant Propagation by Tissue Culture. Part 1: The Technology. Exegetics Ltd., Edington, Wiltshire, England

[28] Al-Khayri J M (2011) Influence of Yeast Extract and Casein Hydrolysate on Callus Multiplication and Somatic Embryogenesis of Date Palm (*Phoenix dactylifera* L.) Scientia horticult. 130(3): 531–535.

[29] Idei S, Kondo K (1998) Effects of NO_3 and BAP on Organogenesis in Tissue Cultured Shoot Primordia Induced from Shoot Apics of *Utricularia praelonga* St. Plant cell rep. 17:451-456

[30] Ramage C M, Williams R R (2002) Mineral Nutrition and Plant Morphogenesis. In vitro cell dev. biol. - plant 38:116-124.

[31] Chi G L, Pua E C, Goh C J (1991) Role of Ethylene on de Novo Shoot Regeneration from Cotyledonary Explants of *Brassica campestris* sp. pekinensis (Lour) Olsson *In Vitro*. Plant physiol. 96:178-183

[32] Beyer E M (1976c) Silver ion: A Potent Anti-Ethylene Agent in Cucumber and Tomato. Hort. scien. 11(3):175-196.

[33] Yang S F (1985) Biosynthesis and Action of Ethylene Hort. scien. 20(1): 41-45.

[34] Bais H P, George J, Ravishankar G A (1999) Influence of Polyamines On Growth Of Hairy Root Cultures of Witloof Chiocory (*Chicorium intybus* L cv Lucknow local) and Formation of Coumarins. J. plant growth regu. 18(1):33-37.

[35] Bais H P, Sudha G, Ravishankar G A (2000b) Putrescine and Silver Nitrate Influence Shoot Multiplication *In Vitro* Flowering and Endogenous Titers of Polyamines in *Cichorium intybus* L cv. Lucknow local. J. plant growth regu. 19:238-248.

[36] Beyer E M (1976a) A Potent Inhibitor of Ethylene Agent in Plants. Plant physiol. 58(3):268-271.

[37] Murashige T, Skoog F (1962) A Revised Medium for Rapid Growth and Bioassays with Tobacco Tissue Cultures. Physiol. plant 15:473–497

[38] Dumas E, Monteuuis O (1995) *In vitro* Rooting of Micropropagated Shoots from Juvenile and Mature *Pinus pinaster* Explants : Influence of Activated Charcoal. Plant cell tiss. org. cult. 40: 231 – 235.

[39] Fraguas C B, Pasqual M, Dutra L F, Cazetta J O (2004) Micropropagation of Fig (*Ficus carica* L.) Roxo de Valinhos Plants. In vitro cell dev. biol.- plant 40:471-474.

[40] Halperin W, Wetherell D F (1964) Adventive Embryony in Tissue Culture of Wild Carrot, *Daucus carota*. Amer. j. bot. 51 : 274 – 283

[41] Ammirato P V (1983) Embryogenesis. In: Evans D A, Sharp W R, Ammirato P V , Yamada Y, editors. Handbook of Plant Cell Culture. Vol. I Macmillan, New York, pp 82–123.

[42] Saxena S, Dhawan V (2001) Large-scale Production of *Anogeissus pendula* and *Anogeissus latifolia* by micropropagation. In vitro cell dev. biol.- plant 37: 586 - 591.

[43] Walia N, Kour A, Babbar S B (2007) An Efficient, *In Vitro* Cyclic Production of Shoots from Adult Trees of *Crataeva nurvala* Buch.Ham. Plant cell rep. 26:277-284.

[44] Chaturvedi R, Razdan M K, Bhojwani S S (2004) *In Vitro* Clonal Propagation of an Adult Tree of Neem (*Azadirachata indica* A. Juss.) by Forced Axillary Branching. Plant sci. 166(2): 501-506.

[45] Chongjin J G, Siewkeng N, Prakash L, Chiangshong, Goh C J, Ng S K, Loh C S (1997) The Role of Ethylene on Direct Shoot Bud Regeneration From Mangosteen (*Garcinia mangostana L.*) Leaves Cultured *In Vitro.* Plant sci. 124:193-202

[46] Kumar S, Sarkar A K, Kuhikannan C (1998) Regeneration of plants from Leaflet Explants of Tissue Culture Raised Safed Siris (*Albizia procera*) Plant cell tiss. org. cult. 54:137-143.

[47] Zhang P, Phansiri S, Puanti-Kaerlas J (2001) Improvement of *Cassava* Shoot Organogenesis by the Use of Silver Nitrate *In Vitro.* Plant cell tiss org cult. 67:47-54

[48] Giridhar P, Reddy O B, Ravishankar GA. (2001) Silver Nitrate Influences *In Vitro* Shoot Multiplication and Root Formation in *Vanilla planifolia* Andr. Curr. sci. 81(9): 1166-1170.

[49] Ganesh S D, Sreenath H L (1996) Silver Nitrate Enhanced Shoot Development in Cultured Apical Shoot Buds of *Coffea arabica* Cv Cauvery (S4347). J. plant. crops 24:577-580.

[50] Fuentes S R L, Calheiros M BP, Manettiflho J, Vieira l G E (2000) The Effects of Silver Nitrate And Different Carbohydrate Sources on Somatic Embryogenesis in *Coffea canephora.* Plant cell tiss. org. cult. 60(1):5-13.

[51] Chi G L, Barfield D G, Sim G E, Pua E C (1990) Effect of $AgNO_3$ and Amino-ethoxyvinylglycine on *In Vitro* Shoot Organogenesis From Seedling Explants of Recalcitrant *Brassica* genotypes. Plant cell rep. 9(4):195-198

[52] Palmer C E (1992) Enhanced Shoot Regeneration from *Brassica campestris* by Silver Nitrate. Plant cell rep. 11(11):541-545.

[53] Vasil I K (1987) Developing Cell and Tissue Culture System for the Improvement of Cereal and Grass Crops J. plant physiol. 128(3):193-218,

[54] Bais H P, Sudha G, Suresh B, Ravishankar G A (2000a) $AgNO_3$ influences *In Vitro* Root Formation in *Decalepis hamiltonii* Curr. sci. 79: 894-898.

[55] Reddy B O, Giridhar P, Ravishankar G A (2001) *In Vitro* Rooting of *Dacalepis hamiltonii* Wight and Arn an Endangered Shrub by Auxins and Root Promoting Agents Curr. sci 81(11):1479-81.

[56] Sunandakumari C, Martin K P, Chithra M, Madhusoodan P V (2004) Silver nitrate Induced Rooting and Flowering *In Vitro* on Rare Rhoeophytic Woody Medicinal Plant *Rotula aquatica* Lour. Ind. j. biotech. 3(3):418-421.

Chapter 11

Production of Useful Secondary Metabolites Through Regulation of Biosynthetic Pathway in Cell and Tissue Suspension Culture of Medicinal Plants

Hu Gaosheng and Jia Jingming

Additional information is available at the end of the chapter

1. Introduction

Medicinal herbs played important roles in human history, from ancient times to now. They have been used for thousands of years to cure diseases, colorize clothes, adjust food taste and keep healthy. It was recorded that, 61% of currently used small molecular drugs are derived from or inspired by natural products from medicinal herbs [1]. However, to cure disease, we need to harvest medicinal herbs, and use part of their tissue for extraction, such as root, leaves, seeds, flowers and so on. That will directly cause problems for the reproduction of these herbs. It has also been noted that the natural habitats have been destroyed due to human activities. With the modern city under construction, the enlarging need for natural resources and the serious pollution, the environment has been never so tough for the growth of medicinal herbs. In another word, the naturally growing medicinal herbs can't fulfill the need of increasing market. Furthermore, the naturally grown medicinal herbs are also different from before, they are carrying more and more herbicide, insecticide and heavy metals, which will cause contamination to the extract and finally cause side effects. Besides, due to the complex structures of secondary metabolites, the chemical synthesis is proved to be cost-inefficient in most cases. Thus, how to produce enough medicinal herbal material in an appropriate manner becomes more and more urgent for the development of pharmaceutical industry all over the world.

In 1934, White proposed the theory of totipotency, and Steward proved the theory in 1952~1953 using carrot cells cultured in liquid media to regenerate whole plant. From then, the cell and tissue culture techniques are developed. As an alternative choice to produce active secondary metabolites, cell and tissue culture of medicinal herbs has obvious advantages:

1. The culture system doesn't need much field which can be used for crop growing;
2. The system is not limited by whether and season changes.
3. The secondary metabolism can be regulated to maximize the production of target compounds.
4. No herbicide and insecticide will be used during the maintaining of the system and therefore, the system is eco friendly.
5. Once the system is established, the content of useful compounds will be more stable than harvested herbs from different areas, which will facilitate the quality control.

Currently, hundreds of medicinal herbs have been used to establish different culture systems, such as callus culture, cell suspension culture, hairy root culture and adventitious root culture. Among these medicinal herbs, endangered species and herbs with positive anti-cancer, neuroprotection, anti-malaria, anti-oxidative activities attracted more attentions. With the development of molecular technology, more and more genetic information related to secondary metabolism becomes available, due to the hard work of generations of scientists from all over the world. After reviewing the research papers about secondary metabolism regulation, there are following strategies being developed to maximize the production of target compounds:

1. Over-expressing the key gene(s) involved in the biosynthetic pathway;
2. Blocking the competitive branches of biosynthesizing target compounds;
3. Blocking the degradation pathways or enhancing the transportation of target compounds;
4. Inhibiting the reproductive growth of plants and increase the biomass of vagetation growth, and increase the production of target compounds.
5. Introduce key genes into microbes and use combinatorial biosynthesis to produce target compounds or important intermediates.

Furthermore, plant cell culture techniques are playing more and more important roles in confirmation of gene function, expression, and contribution in the biosynthetic pathways of secondary metabolites. Plant cell and hairy root culture have become powerful tools in the genetic research fields. The reason is mainly because of the controlled growing environments (inoculated cells, light, pH, nutrient, shaking speed, temperature, treatments and pathogen-free etc.), stability and reproducibility of culture cells or organs.

Until now, many plant species have been used to establish different culture system, and useful compounds been targeted, among which, the accumulation of over 30 target compounds exceeded the content in wild plants. However, there are only shikonin (12% of dry weight) [2], ginsenoside (in 20000 L scale) [3], toxol (in 75000 L scale) [4] and berberine (13.2% of dry weight) [5] being produced using plant cell and tissue culture techniques in application scale, which is quite embarrassing. The bottle neck limiting the application of plant cell and tissue culture lies in the shortness of this technique: long culture period, low yield, high cost etc. However, there is another factor to be considered, that, in most medicinal plant, even the single compounds are being identified, few of them have potent activity against serious diseases, besides, in oriental medicines, most medicinal herbs are boiled in water together, following various combinations, which made finding the effective compounds more difficult, and single compound(s) can't represent the curing effects. When we analyze the cultured callus, cells or tissues, we always can find the differences, compared with wild plant, which made it difficult to use the cultured cells or organs directly as wild plant materials. In one word, only if the target compound has high medicinal value, trying to increase the accumulation in culture system or combinatory biosynthesis microbes and finally purify the compound is worthy of the efforts. Without high medicinal value, the research can only be considered as a basic research solving mechanism related puzzles. Recently, combination of treatments are being developed in order to achieve high yield final products, such as elicitor treatment, repeated elicitor treatment, precursor feeding, over expression of key genes. In the future, the plant cell and organ culture system might have more potential in pharmaceutical industries.

In the following contents, the establishment of different culture system will be introduced and recent progress of secondary metabolism regulation will be reviewed and discussed.

2. Establishment of culture systems

To regulate the secondary metabolism of cells aiming at achieving maximized production of useful compounds, culture system should be established first. Here, establishment of three main culture systems will be introduced: cell suspension culture, hairy root culture and adventitious root culture. Three kinds of system are all liquid-form culture systems. The liquid form culture has advantages compared with solid form culture. Firstly, the contacting surface area of cells and media is much larger, and therefore, the nutrient will be much easier to be utilized by cells with higher efficiency. Secondly, compounds harmful for cell growth formed during the culture can be effectively diluted and avoid the inhibition of cell growth. Thirdly, in liquid culture system, the media is well mixed and dissolved oxygen can be monitored and controlled, finally, the scale-up of culture system is easier with the development of fermentation techniques.

2.1. Establishment of cell suspension culture.

To establish cell suspension culture of a medicinal herb, firstly, the callus should be induced. During the callus induction, explants of plant origin should be surface sterilized and sliced into pieces about 0.5 cm^3 in clean bench, and inoculated on autoclaved solid basic media (MS, B5, N6, White etc.) supplemented with sucrose, hormones and agar. After the callus was induced, it should be sub-cultured. After sub-culture, usually, various types of callus can be found with different texture and color. Callus of various types should be introduced into liquid media (most of the time, after removal of agar, the formula of solid media can be used for liquid media preparation). Contents determination of active compounds should be carried out for cell line selection. In this step, cultured cells of different types should be sampled at each growth stage, and the cell samples will be subjected to biomass measurement and content determination as shown in Fig. 1, because cell of different types might be growing at different speed, and the maximal yield of target compounds can also be different when collected at same culture time.

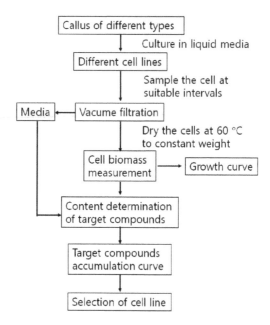

Figure 1. Flow chart of cell line selection

Once the cell line is selected, investigation on cell growth, cell viability, pH changing, carbon source consumption, enzyme activity, gene expression, target compounds accumulation can be carried out following the flow chat shown in Fig. 2. The total dry weight (DW) is calculated following the equation: $DW=W4\times(W1+W2+W3)/W3$. To be noticed is that, the DW is a very important index, which will be used in growth curve, target compound yield, enzyme activity curve, gene expression curve preparations. In some cases, target compounds can be secreted into media, and therefore, the content determination of target compounds in liquid media should also be taken into consideration. The compounds detected in media and in dried cells should be summed to represent the compound producing ability.

2.2. Establishment of hairy root suspension culture.

For many medicinal herbs, roots are the medicinal part used for extraction, such as *Panax giseng, Panax notogiseng, Coptis chinensis, Savia miltiorrhiza* and so on. Furthermore, it has also been found that, in differentiated tissues, secondary metabolites accumulation is usually higher than callus and cells without differentiation. Therefore, hairy root can be an effective candidate culture form.

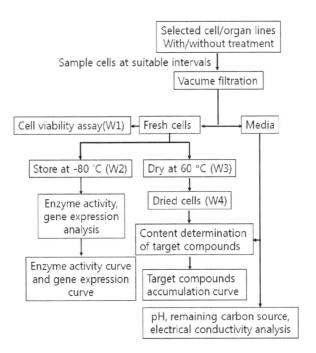

Figure 2. Flow chart of different investigation using cell sample and media

Agrobacterium rhizogenes, belonging to the family of Rhizobitaceae and genus of *Agrobacterium*, is a gram-negative soil bacterium. It has flagellum and can swim. The victim host can be most dicotylendous plants, few monocotylendous and specific gymnosperms plants, hairy root will be formed on the infected sites.

It has been demonstrated that, outside of chromosome, huge covalent circular double stranded DNA exists, and the size is between 180 and 250 kb, which is named *Ri* (Root inducing) plasmid. *Ri* plasmid has two major parts, virulent region (*Vir*) and transfer region (T-DNA). With the assistance of *Vir* region, T-DNA can be transferred to host plant, and integrated to host plant chromosome. In T-DNA, there is a gene called opine synthase with a eukaryotic promoter. After the integration happens, opine synthase promoter started to work and plant cell will be forced to synthesize a specific non-coding amino acid, opine, which will be used as the only carbon source for infected cells. The infected cells grow vigorously, and the phenotype is hairy root. Following are hairy root induction protocol:

1. Preparation of sterile explants

Explants can be obtained from seedlings germinated from sterile seeds of medicinal herbs. Young tissues (leaf, stem, shoot) of plants collected from field can also be used as explants after surface sterilization. After sterilization, the explants can be inoculated on basic solid media supplemented with (50-200mM) acetosyringone.

2. Preparation of *A. rhizogenes*

Bacterium stock stored at -80°C deep freezer can be striked on solid YEP media, and single colony is selected for culture in liquid YEP media for 16 h in dark. Culture condition is 28°C and shaking speed is 200rpm. After the OD_{600} reached 0.6-1.0, the bacterium culture can be used for infection.

3. Infection

Centrifuge the bacterium culture obtained above at 4000 rpm for 30 min at 4 °C, and resuspend the culture in same volume YEP media supplemented with 50-200uM acetosyringone. Immerge the wound sterile explants in resuspended Agrobacterium culture for 10 min and dry its surface on filter paper to remove excess agrobacterium. The wound is essential for successful hairy root infection, and usually there are a few wounding methods using sterile needles and sharp knife. The wound doesn't need to be serious, slight wound exposing inside tissue would be enough for infection to happen. Inoculate the infected tissue in basic media supplemented with acetosyringone.

4. Co-culture

Incubate the petri-dish in dark at 20-22°C for 2-5 days until obvious bacterium growing is visible.

5. Wash

Wash the infected explants from step 4 with 75mg/L hygromycin and carbonicillin water solution for at least 5 times. Vortex the tissue to remove the excess bacterium.

6. Hairy root growing

After washing, the tissue can be inoculated on solid media with acetosyringone and carbonicillin. It will take about two weeks for the hairy root growing until visible. During this process, once agrobacterium colony is visible on solid media, the tissue should be washed again following step 4.

7. Hairy root culture

Once hairy root is germinated from infected sites, the root can be separated from explants and culture on solid media until enough biomass is obtained. Then the hairy root will be inoculated in liquid media incubated in shaking incubator at 25 °C and at about 130 rpm.

2.3. Establishment of adventitious root suspension culture.

Adventitious root is another model culture system, which can produce identical components compared with wild growing plants. Compared with hairy root culture, adventitious root will not produce opine, grow slower and require no genetic transformation techniques. Therefore, for the directly further use of cultured roots, this system provides safer product, after all, opine is a harmful compound for human beings. Adventitious root can be induced from sterile explants, callus and seedlings of plant. The hormone used most widely is IBA. After induction, the callus or explants together with adventitious root can be inoculated into liquid media and cultured in shaking flask or bioreactors. Until now, the most famous application is the adventitious root culture of *Panax giseng*, that, the culture system has been scaled up to 10 tons and the total saponin yield reached 500mg/L/day [6].

2.4. Application of bioreactors.

The application of bioreactor is one of the prerequisites for industrialization of plant cell culture. Therefore, bioreactor design specific for plant cell is important for the production of secondary metabolites using cell culture techniques. Based on the structure, plant cell bioreactor can mainly be divided as 5 types: mechanical stirring, air-lifting, bubbling, nutrient mist and temporary immersion bioreactors. These bioreactors have some characteristics in common that they can provide well mixed media and sterile air. Different from shaking flask, that the shaking will allow the media contact with air in the flask continuously, the media in bioreactor should be stirred or lifted by air bubbles, or mixed with air and then provide to cultured cells or tissues. Therefore, provision of air becomes very important for bioreactors. In bioreactors, there are several sensors monitoring the pH changes, temperature, dissolved oxygen (DO) and generated bubbles, the automated bioreactor can control these indexes following the order set by the operators.

Usually, the mechanical stirring bioreactor can produce highest DO index, however, it is not used widely in plant cell and tissue culture, because they are sensitive to shear force. In case of air-lift and bubbling bioreactors, they have similar structures and used quite often in plant cell and tissue culture. In hairy root culture, the bioreactors are always equipped with layers of meshes made of stainless steel, to provide attaching area for hairy roots. The nutrient mist and temporary immersion bioreactors have similar characteristics and they are both used for tissue culture, such like hairy root, adventitious root, shoots, bud clusters. They are all composed of two parts, media storage part and tissue culture part. After mixing of media and sterile air, for nutrient mist bioreactor, the media is sprayed through atomizer into tiny droplets on the surface of cultured tissues, and for temporary immersion bioreactor, the media is pumped to the culture part and maintain for short time and then the media is recovered to storage tank.

There are successful examples for cell and tissue culture of medicinal herbs to produce useful secondary metabolites. *Digitalis purpurea* cell line was cultured in airlift bioreactors, and the yield of methyl isopropyl hydroxyl digitoxin reached 430 mg/L; *Phalaenopsis aphrodita* protocorm like body culture in temporary immersion bioreactor for micro-propagation [7]; *Panax ginseng* adventitious root culture in bollon type bubbling bioreactor at scale of 20 tons [3]. All these achievement demonstrated the industrial potential of the application of bioreactor in producing secondary metabolites with high medicinal value.

3. Biosynthetic pathways of secondary metabolites

3.1. Biosynthetic pathway introduction of secondary metabolites in medicinal plants

During long period evolution, plants struggling to survive gradually gain the ability to synthesize various kinds of secondary metabolites with bioactivities. These compounds played important role in defencing insects, herbivores, microbial pathogens, competing with other plants, and faciliting pollination and reproduction. Based on the structures, the second metabolites can be classified into alkaloids, flavonoids, phenylpropanoids, quinones, terpenoids, steroids, tannins and proteins. These compounds are biosynthesized through series enzyme catalyzed reactions using simple building blocks in different ways. There are several main biosynthetic pathways in plants, including shikimic acid pathway (phenylpropanoids), movalonic acid pathway (quinones), 2-C-methyl-D-erythritol-4-phosphate pathway (quinones), amino acid pathway (alkaloids), acetate-malonate pathway (fatty acid, phenols and quinones) and combined pathways (flavonoids). These biosynthetic pathways are well reviewed and discussed in publication [8], and the current determined steps in biosynthesis pathways are also centralized and documented in KEGG [9] website, therefore, won't be further introduced in this chapter.

3.2. Molecular cloning and characterization of key enzymes involved in specific biosynthetic pathway.

Gene cloning, transformation and regulation have achieved significant progress in biosynthetic pathway of secondary metabolites. The biosynthetic pathway elucidation (Fig. 3) and gene regulation research of vinblastin and vincristine, anticancer compounds from *Catharanthus roseus*, is a representative example. Due to the low contents of these two compounds in naturally growing plant, it becomes important to clone and regulate expression of key genes (DXR, SLS, G10H, STR) involved in the pathway, in order to achieve high yield of these two compounds in invitro culture systems, which will be based on well understanding of terpenoid indole alkaloids biosynthesis in *C. roseus* [10].

As to the approaches applied in the cloning of genes involved in secondary metabolites biosynthetic pathway, there are methods based on Polymerase chain reaction (PCR) and library construction. The former methods include Rapid amplification of cDNA ends PCR (RACE PCR) and RT-PCR. However, these methods need at least partial gene sequence information, or gene sequence from other plant species to synthesize degenerate primers. If there are no reported reference sequences, these methods will not be able to work well. Under this situation, library based method can be selected for gene cloning, however, the library based methods are non specific enough, like cDNA library, BAC library. That's because of the low ratio of genes involved in secondary metabolism in total expressed mRNAs. As the progress of genomics, functional genomics provided us useful methods in cloning the specific genes involved in the secondary metabolism, such as Subtractive hybridization 'Differential screening' Microarray assay and Serial Analysis of Gene

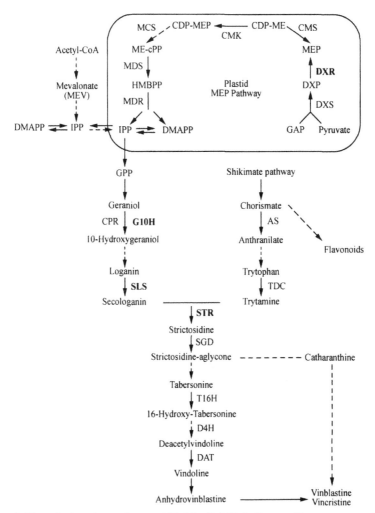

Figure 3. Biosynthetic pathway of terpenoid indole alkaloids in *C. roseus* (Key enzymes are shown in bold)

Expression. Functional genomics focused on the selection of treatment responsive genes and to clone genes involved in the biosynthetic pathways, the treatments need to be specific enough to result in the improvement of target secondary metabolites.

3.3. Elicitors and signaling pathways

Elicitors are substances that can trigger the hypersensitive reaction in treated plant cells. Due to the effective up-regulation of genes expression, and further activation of secondary metabolism, and the improvement effects of secondary metabolites accumulation, elicitors are used widely in medicinal plant cell and tissue culture to maximize the production of target compounds.

Elicitors can be divided into biotic and abiotic elicitors. The former one includes fugal polysaccharides, proteins, cell debris and conidium. These kinds of elicitors are made from cultured fungi and some species are specific for specific kinds of secondary metabolites accumulation, for example, *Armillaria mellen* elicitor for alkaloids accumulation; *Verticillium dahlia* for phenolic compounds accumulation; *Botrytis sp* for terpenoids accumulation, oligosaccharides for saponins accumulation and yeast extract for flavonoids accumulation. As to abiotic elicitors, heavy metals ions, rare metal ions, UV lights, osmotic stress and even sonication have all been reported to have positive effects towards improvement of secondary metabolites. Comparing the application potential, biotic elicitor attracted more attention due to their advantages of low cost, little side effects, strong elicitation effects and easy manipulation.

When using elicitors, the main optimizing characters are treatment time, concentration and selection of elicitor kinds. In plant cell/organ culture, it is believed that, treat cells/organs with elicitors at the stabilization stage can increase the accumulation of secondary metabolites significantly. However, there are also reports that treat cells at the beginning of culture, which also improved the final product yield remarkably. In this case, there is possibility identifying key intermediates involved in the biosynthetic pathway. As to the concentration, too high concentration treatment can result in decrease of biomass accumulation and even the content of secondary metabolites can be improved, however, the total yield will still not be maximized. Therefore, to use an elicitor, the conflict between biomass accumulation and secondary metabolite accumulation should be balanced and the optimization should be carried out. Recently, a strategy of repeated elicitation has been developed, and the treatment can result in the increase of target compounds accumulation as well as related enzyme activity and gene expression more than one time, and therefore, attracts attention in the field of secondary metabolism regulation aiming at maximizing the yield of compound of interest.

3.4. Precursor feeding

Precursors are compounds existed in upstream of target compounds in biosynthetic pathway. Most intermediates can be used as precursors. Upstream precursors are converted into down-stream compounds after specific enzyme catalysis, the concentration of precursors determines the reaction speed. At higher concentration, the reaction speed is usually higher than that when precursor concentration is lower. To improve the yield of secondary metabolites in plant cell and tissue culture, precursor feeding is an effective approach. However, precursor can inhibit cell growth and enzyme activity if concentration is too high. In some cases, high concentration of intermediates can even be toxic for cells. Originally, in cells, these intermediates are all maintained in very low concentration, and converted to down-stream compounds quickly. Therefore, for one specific plant species, precursor concentration should be carefully adjusted, compared and optimized.

The precursor feeding experiment can also be used as a powerful tool for the elucidation of biosynthetic pathways. Isotope labeled precursors can be fed in the media, and as the cell growing and biosynthesis going on, secondary metabolites synthesized can be isotope labeled at different positions. After harvesting the cells, purifying the compounds and elucidating their structures, the biosynthetic origin can be concluded. This method is classic but also most authentic method in elucidating the biosynthetic origin of a specific compound. However, the disadvantages are also obvious, first of all, the precursors are isotope labeled and is dangerous to handle; secondly, the purification of isotope labeled end product requires lots of cells; thirdly, some intermediates are maintained in very low concentration in living plant cells, which will makes the purification of these intermediates very difficult. Therefore, if the isotope labeled precursor feeding combine with elicitor treatment or specific enzyme inhibitors treatment, the problems might be solved.

3.5. Application of specific enzyme inhibitors in regulating the biosynthetic pathway

In plant cells, all secondary metabolites are derived from glucose that is biosynthesized during photosynthesis. In most medicinal plants, and even model plants, there are several kinds of secondary metabolites accumulated. For example, there are about 15 flavonoids, 50 terpenoids, 15 fatty acid, 20 phenylpropanoids, and 35 glucosinolates in *A. thaliana* [11].

Different kinds of compounds are competitively related with each other. To improve the accumulation of one specific kind of compounds or blocking the other main biosynthetic branches, genetic transformation method is ideal for the purpose. However, the transformation method of most medicinal plants has not been established. On the other hand, elicitor treatment is not specific, even the treatment can result in the improvement of secondary metabolites accumulation. Therefore, both methods can't provide specific evidence for contribution of single gene in plant secondary metabolites biosynthetic pathway. Under these circumstances, how to study the contribution of specific gene(s) involved in the biosynthesis pathway to the production of target compounds?

If we treat the cells with enzyme specific inhibitors, the enzyme activity will be specifically inhibited, and result in the corresponding consequences. After content determination of target compounds and enzyme activity assay, correlation analysis between enzyme activity and target compounds accumulation can be carried out, and therefore get to know the contribution of specific enzyme in the biosynthetic pathway of target compounds. This method can also be used to identify key enzymes or rate-limiting steps in the pathway. The application of inhibitors can also be used for other purposes. When treated with specific inhibitors blocking the competitive branch, more carbon flux will be channeled to the biosynthesis of target compounds, and result in the improved accumulation. Furthermore, if the cells are treated with specific inhibitors, the enzyme activity will be inhibited, and the upstream compounds can be consequently accumulated at higher concentration which will facilitate the identification and characterization of key intermediates.

3.6. Application of genetic techniques in regulating the biosynthetic pathway

With the development of molecular technology, more and more genetic information becomes available online. Until now, there are dozens of plant species whose genome has been sequenced, such like *Arabidopsis thianalia*, *Oryza sativa*, *Artemisia annua* and so on. Besides, genetic techniques (EST, RACE PCR, transformation, T-DNA tagging, siRNA, miRNA etc.) have also been carried out in other medicinal plants, like *Panax giseng*, *Saussurea involucrate* and so on. These techniques provided tons of information and possibilities for the manipulation of target compound biosynthesis. Japanese scientists introduced a flavonoid 3'5' hydroxylase gene into rose plant with red flowers, and the transgenic plant showed blue-purple flowers [12]. In *Saussurea involucrate* hairy root culture, a chalcone isomaerase gene was introduced and over expressed, which resulted in 12 times production of apigenin of that in control hairy root [13]. Tyrosine decarboxylase gene isolated from parsley, was introduced in potato, and in transgenic plant, salidroside was detected [14]. Salidroside is a phenylethanoid compound with significant adaptive effects and is the main active compound in *Rhodiola sachalinensis*. This report showed us the potential of producing new pharmaceutically important compounds besides original medicinal herb.

3.7. Combinatorial Biosynthesis in production of useful secondary metabolites

Combinatory synthesis is a method to combine metabolism pathways of different species genetically. As the development of molecular biology, and the proposal of creative experimental methods and idea, the concept of combinatory will be changed in the future, but there is no doubt about its importance in post genomics.

Currently, most research related to combinationary synthesis are using two kind of microbes in their experiment, *Escherichia coli* and *Saccharomyces cerevisiae*. Two microbes are not accidentally selected as model species. Experimental skills related to their mutation, culture, transformation and expression are all well developed. Besides, the whole genome sequence information is also available which provided detailed information for the manipulation and genetic engineering of two microbes.

To be noticed is that, full biosynthetic pathways of secondary metabolites are not involved in the microbes or at least partially not involved. Therefore, to produce the specific compound in microbes, whole biosynthetic pathway or part of the pathway must be introduced into the target microbes (through high copy number plasmid or integration to chromosome). This is the main idea of using *E. coli* and *S. cerevisiae* as the vehicles of combinatory biosynthesis, and also the bottle-neck limiting the application of this technique, due to the poor understanding of biosynthetic pathway of secondary metabolites in plants. Thus, currently, most researchers are focusing on high-yield intermediate production and use chemical semi-synthesis to produce the final product of interest or precursor feeding in the cultured transformed microbes.

1. Anthocyanin production in genetic engineered *E. coli*

Due to the wide application of anthocyanin in naturally occurring dye, edible pigment, and antioxidant activity, its biosynthetic pathway has attracted attention of scientists and lots of important experimental experiences and results are reported. Its biosynthetic pathway is one of the best understood pathway of natural products. In this study, genes of flavanone 3′ hydroxylase (*F3′H*), anthocyanidin synthase (*ANS*) cloned from *Malus domestica* and genes cloned from *Anthurium andraeanum*, dihydroflavonol 4-reductase (*DFR*) and UDP-glucose:flavonoid -3-O- glucosyltrasferase (*3-GT*) are constructed in one artificial gene cluster and expressed in *E.coli*. In E. coli culture media, when fed the precursor of anthocyanin biosynthesis, the naringenin, or eriodictyol, anthocyanin production was detected in culture media [15].

2. Artemisic acid production in engineered *S. cerevisiae*

Artemisine is the most effective anti-malaria compound until now, traditionally, the compound is purified from dry harvested aerial part of *Artemisia annua*. The purification is time and labor consuming, and chemical synthesis has been proved cost unattractive. Therefore, genetic engineering of yeast cell is used to produce the precursor of this compound, artemisinic acid, and semi synthesis is used to further synthesize the artemisine. Genes isolated from *A. annual*, amorphadiene synthase (*ADS*) and cytochrome P450 monooxygenase (*CYP71AV1*) were interegrated into yease chromosome. Furthermore, the biosynthesis pathway of farnesyl pyrophosphate (*FPP*) was also engineered to produce more FPP, and the useage of FPP to steroid was also blocked to increase the FPP flux into artemisinic acid. The engineered yeast cell was cultured in YEP media, and the yield of artemisinic acid reached 115 mg/L. besides, the purification process is also very simple that, the final product was all on the surface of yeast cells, and by centrifugation and acidic buffer washing, 95% of artemisinic acid produced can be recovered [16].

4. Conclusions

Plant cell and organ culture is fast developing in the field of secondary metabolism regulation, with the development of molecular biology, deepen understanding of the biosynthetic pathways of natural products and newly developed treatment strategies. It is now not only an approach to produce target compounds, but also powerful tool for gene function research in post-genomic era. More and more medicinal herbs have been used to establish different culture systems, which will facilitate preserving the natural resources and improve the possibilities of producing compounds of interest at industrial level. In order to achieve target compounds yield high enough for industrial application, or to elucidate the full biosynthetic pathway, methods such as key gene(s) transformation, elicitor treatment, precursor feeding, inhibitor treatment should be combined in an appropriate manner. However, due to the culture characteristics and final product recovery process, combinatorial biosynthesis showed even more potential in the industrial application.

Author details

Hu Gaosheng and Jia Jingming
School of Traditional Chinese Material Medica, Shenyang Pharmaceutical University, Shenyang, Liaoning, P. R. China

5. References

[1] Newman D J., Cragg G M., & Snader K M. Natural Products as Sources of New Drugs over the Period 1981-2002. *J. Nat. Prod.* 2003, *66*, 1022-1037.

[2] Tabat A M & Fujit A Y. Product ion of shikon in by plant cultures [M]. Biotechnology in Plant Science. Academica Press, 1985: 207-218.

[3] Paek K Y, Murthy H N, Hahn EJ, Zhong JJ Large scale culture of ginseng adventitious roots for production of ginsenosides. *Adv Biochem Eng Biotechnol.* 2009, 113:151-76.

[4] O. Expósito1, M. Bonfill1, E. Moyano, *et al.* Biotechnological Production of Taxol and Related Taxoids: Current State and Prospects. *Anti-Cancer Agent Med Chem*, 2009, 9:109-121.

[5] Sato, F. & Yamada, Y. High berberine-producing cultures of *Coptis japonica* cells *Phytochem.* 1984, 23:281.

[6] Paek KY, Hahn EJ & Son SH. Application of bioreactors of large scale micropropagation systems of plants. *In vitro Cell Dev Biol-Plant.* 2001, 37:149-157.

[7] Park SY, Murthy HN & Paek KY. Mass multiplication of protocorm like bodies using bioreactor system and subsequent plant regeneration in *Phalaenopsis. Plant Cell, Tiss Org Cult.* 2000, 63: 67-72.

[8] Paul M. Medicinal Natural Products: A Biosynthetic Approach. John Wiley & Sonseds. 2nd edition, 2001

[9] http://www.genome.jp/kegg/

[10] Han Mei. Cloning and expression of cDNA encoding key enzymes (DXR, SLS, G10H, STR) in terpenoid indole alkaloids biosynthetic pathway from *Cathrarnthus roseus. Plant Research.* 2007, 27(5):564-568.

[11] John C D'Auria. The secondary metabolism of Arabidopsis thaliana: growing like a weed. *Curr Opin Plant Biol*, 2005, 8:308-316.

[12] Yukihisa K., Masako F. M., Yuko F. et al. Engineering of the Rose Flavonoid Biosynthetic Pathway Successfully Generated Blue-Hued Flowers Accumulating Delphinidin. *Plant Cell Physiol.* 2007, 48(11): 1589-1600.

[13] Feng-Xia Li, Zhi-Ping Jin, De-Xiu Zhao et al. Overexpression of the *Saussurea medusa* chalcone isomerase gene in *S. involucrata* hairy root cultures enhances their biosynthesis of apigenin. *Phytochem*, 2006, 67:553-560.

[14] Jorn L. Accumulation of tyrosol glucoside in transgenic potato plants expressing a parsley tyrosine decarboxylase. *Phytochem*, 2002, 60:683

[15] Yan Y. J. Metabolic Engineering of Anthocyanin Biosynthesis in *Escherichia coli. Appl Environ Microb*, 2005, 71 (7):3617-3623.

[16] Ro D K, Paradise E M, Ouellet M, *et al.* Production of the antimalarial drug precursor artemisinic acid in engineered yeast. *Nature*, 2006, 440(13):940-943.

Permissions

The contributors of this book come from diverse backgrounds, making this book a truly international effort. This book will bring forth new frontiers with its revolutionizing research information and detailed analysis of the nascent developments around the world.

We would like to thank Dr. Annarita Leva and Dr. Laura M. R. Rinaldi, for lending their expertise to make the book truly unique. They have played a crucial role in the development of this book. Without their invaluable contribution this book wouldn't have been possible. They have made vital efforts to compile up to date information on the varied aspects of this subject to make this book a valuable addition to the collection of many professionals and students.

This book was conceptualized with the vision of imparting up-to-date information and advanced data in this field. To ensure the same, a matchless editorial board was set up. Every individual on the board went through rigorous rounds of assessment to prove their worth. After which they invested a large part of their time researching and compiling the most relevant data for our readers. Conferences and sessions were held from time to time between the editorial board and the contributing authors to present the data in the most comprehensible form. The editorial team has worked tirelessly to provide valuable and valid information to help people across the globe.

Every chapter published in this book has been scrutinized by our experts. Their significance has been extensively debated. The topics covered herein carry significant findings which will fuel the growth of the discipline. They may even be implemented as practical applications or may be referred to as a beginning point for another development. Chapters in this book were first published by InTech; hereby published with permission under the Creative Commons Attribution License or equivalent.

The editorial board has been involved in producing this book since its inception. They have spent rigorous hours researching and exploring the diverse topics which have resulted in the successful publishing of this book. They have passed on their knowledge of decades through this book. To expedite this challenging task, the publisher supported the team at every step. A small team of assistant editors was also appointed to further simplify the editing procedure and attain best results for the readers.

Our editorial team has been hand-picked from every corner of the world. Their multi-ethnicity adds dynamic inputs to the discussions which result in innovative outcomes. These outcomes are then further discussed with the researchers and contributors who give their valuable feedback and opinion regarding the same. The feedback is then collaborated with the researches and they are edited in a comprehensive manner to aid the understanding of the subject.

Apart from the editorial board, the designing team has also invested a significant amount of their time in understanding the subject and creating the most relevant covers. They scrutinized every image to scout for the most suitable representation of the subject and create an appropriate cover for the book.

The publishing team has been involved in this book since its early stages. They were actively engaged in every process, be it collecting the data, connecting with the contributors or procuring relevant information. The team has been an ardent support to the editorial, designing and production team. Their endless efforts to recruit the best for this project, has resulted in the accomplishment of this book. They are a veteran in the field of academics and their pool of knowledge is as vast as their experience in printing. Their expertise and guidance has proved useful at every step. Their uncompromising quality standards have made this book an exceptional effort. Their encouragement from time to time has been an inspiration for everyone.

The publisher and the editorial board hope that this book will prove to be a valuable piece of knowledge for researchers, students, practitioners and scholars across the globe.

List of Contributors

Altaf Hussain
Qarshi University, Lahore, Pakistan

Iqbal Ahmed Qarshi
Qarshi Industries (Pvt.) Ltd. Lahore, Pakistan

Hummera Nazir and Ikram Ullah
Plant tissue Culture Lab, Qarshi Herb Research Center, Qarshi Industries (Pvt.) Ltd. Hattar, Distt. Haripur, KPK, Pakistan

Abobkar I. M. Saad and Ahmed M. Elshahed
Department of Botany and Microbiology, Faculty of Science, Sebha University, Libya

Mustafa Yildiz
Department of Field Crops, Faculty of Agriculture, University of Ankara, Diskapi, Ankara, Turkey

Rosa Mª Pérez-Clemente and Aurelio Gómez-Cadenas
Department of Agricultural Sciences, Universidad Jaume I. Campus Riu Sec. Castellón, Spain

S.E. Aladele, A.U. Okere, E. Jamaldinne, P.T. Lyam and O. Fajimi
Biotechnology Unit, National Centre for Genetic Resources and Biotechnology (NACGR-AB), Moor Plantation, Ibadan, Oyo State, Nigeria

A. Adegeye
Department of Forest Resources Management, University of Ibadan, Ibadan, Oyo State, Nigeria

C.M. Zayas
Genetic and Biotechnology Department, National Institute for Sugarcane Research, Boyeros, Havana City, Cuba

A.R. Leva, R. Petruccelli and L.M.R. Rinaldi
CNR IVALSA Trees and Timber Institute, Firenze, Italy

Moacir Pasqual, Leila Aparecida Salles Pio, Ana Catarina Lima Oliveira and Joyce Dória Rodrigues Soares
Federal University of Lavras (UFLA), Department of Agriculture, Lavras, MG, Brazil

Giuseppina Pace Pereira Lima
Institute of Biosciences, São Paulo State University, Botucatu, São Paulo, Brazil

Renê Arnoux da Silva Campos
Mato Grosso State University, Cáceres, Mato Grosso, Brazil

Lilia Gomes Willadino
Federal Rural University of Pernambuco, Recife, Pernambuco, Brazil

Terezinha J.R. Câmara
Federal Rural University of Pernambuco, Recife, Pernambuco, Brazil

Fabio Vianello
University of Padua, Padova, Italia

Moacir Pasqual, Joyce Dória Rodrigues Soares and Filipe Almendagna Rodrigues
Federal University of Lavras (UFLA), Department of Agriculture, Lavras/MG, Brasil

Edvan Alves Chagas
Brazilian Agricultural Research Corporation (EMBRAPA), Distrito industrial, Boa Vista-RR, Brasil

Y.K. Bansal and Mamta Gokhale
Plant Tissue Culture Laboratory, Department of Biological Science, R.D. University, Jabalpur-M.P. India

Hu Gaosheng and Jia Jingming
School of Traditional Chinese Material Medica, Shenyang Pharmaceutical University, Shenyang, Liaoning, P. R. China